D1300046

WILDLIFE DAMAGE CONTROL

WILDLIFE DAMAGE CONTROL

JIM HONE

National Library of Australia Cataloguing-in-Publication entry

Hone, Jim.
Wildlife damage control.

Bibliography.
Includes index.
ISBN 978 0 643069 59 6.

1. Vertebrate pests – Control. I. Title.

632.66

Published exclusively in Australia, New Zealand and Asia, and non-exclusively in other territories of the world (excluding Europe, the Middle East, Africa and North America), by:

CSIRO PUBLISHING
150 Oxford Street (PO Box 1139)
Collingwood VIC 3066
Australia

Tel: 03 9662 7666 Int: +61 3 9662 7666
Fax: 03 9662 7555 Int: +61 3 9662 7555
Local call: 1300 788 000 (Australia only)
Email: publishing.sales@csiro.au
Website: www.publish.csiro.au

Published exclusively in Europe, the Middle East, Africa and North America and non-exclusively in other territories of the world (excluding Australia, New Zealand and Asia), by CABI (CABI is a trading name of CAB International), with ISBN 978 1 84593 245 9.

CABI
Wallingford, Oxfordshire
OX10 8DE
United Kingdom

Tel: 01491 832 111 Int: +44 1491 832 111
Fax: 01491 829 292 Int: +44 1491 829 292
Email: publishing@cabi.org
Website: www.cabi.org

Cover: image by istockphoto

Set in Adobe Plantin, Minion and Stone Sans
Edited by Adrienne de Kretser
Cover and text design by James Kelly
Typeset by Desktop Concepts Pty Ltd, Melbourne
Index by Russell Brooks
Printed in Australia by Ligare

Contents

Preface

Wildlife causes many problems worldwide. Solutions to those problems could be discovered separately in each location for each vertebrate species, or they may be found by identifying and using common principles in the assessment and control of pest damage. The latter approach is the theme of this book.

Wildlife Damage Control focuses on discussion of principles, not on the pest species per se. The list of principles for damage control should not be interpreted as complete – readers can probably suggest additional principles for evaluation, and their contribution will be welcome. Such a process will contribute to the further development of a rigorous science of wildlife damage control and the application of more reliable knowledge in management.

The book is written for use by scientists in wildlife management, environmental science and natural resource management, and for final-year undergraduate and graduate students. It will also be relevant for wildlife managers around the world and anybody else involved or interested in control of wildlife damage. Such control may occur for biodiversity conservation, in production industries such as agriculture, forestry and fisheries, in human and animal health and safety, and in urban areas of wildlife conflict such as golf courses, parks and gardens. *Wildlife Damage Control* assumes some knowledge of ecology, conservation biology, agriculture, statistics and mathematics. Comments and feedback on the book are welcome.

The book represents the results of work of many scientists from around the world. There is a vast literature worthy of inclusion but space does not permit reviewing all of that literature. A lack of inclusion of one's research results is not a reflection on the quality of that research.

I thank the University of Canberra, CSIRO Sustainable Ecosystems and Manaaki-Whenua Landcare Research (New Zealand) for support and facilities, and students at the University of Canberra and staff at Manaaki-Whenua Landcare Research for feedback on draft versions. Special thanks to P. Caley for useful comments on draft chapters, and to A. Crabb and A. de Kretser for editorial inputs.

1 Introduction

Background

Wildlife may cause problems in conservation, agriculture, forestry and fisheries, to human and animal health and safety, to recreation areas and in cities and towns. For example, *Rattus* species introduced onto oceanic islands can be important predators of seabirds and their eggs. Predatory birds, such as harriers, can kill other birds that are managed for hunting or conservation. Carnivores such as wolves (*Canis lupus*), coyotes (*Canis latrans*), foxes (*Vulpes vulpes*) and lynx (*Lynx lynx*) can prey on livestock. Badgers (*Meles meles*) and brushtail possums (*Trichosurus vulpecula*) are important hosts of bovine tuberculosis in Britain and New Zealand respectively. House mice (*Mus domesticus*) damage crops in Australia and other rodents cause substantial crop damage in south-east Asia. *Quelea* finches damage crops in parts of Africa, as do elephants (*Loxodonta africana*). Pigeons (*Columba livia*) foul statues in urban areas and geese foul the greens of golf courses. Australian magpies (*Gymnorhina tibicen*) attack people in urban areas.

Forms of wildlife damage are obviously many and varied. They can also be costly. Conover *et al.* (1995) costed wildlife damage in the US as approaching US$3 billion. Subsequent estimates of wildlife damage in the US put the cost at approximately US$22 billion annually (Conover 2002).

If pests are defined as species that have undesirable effects, wildlife can at times be considered pests. Obviously they have other values also, such as utilitarian value when people use them for food, great scientific value as subjects of basic ecological research, and aesthetic value. However, this book focuses on the pest aspects of wildlife. This field of science is known by different titles around the world. In the US it is known as 'animal damage control', 'human–wildlife conflicts' or 'wildlife damage'. 'Vertebrate pest control' is the preferred term in Europe, Asia, Africa, Australia and New Zealand, and in many places the term is 'overabundance'.

Control has traditionally been by lethal methods such as trapping, shooting and poisoning, or non-lethal methods such as fencing, repellents and habitat manipulation. Recently there has been increasing interest in controlling damage by reducing fertility of pests. This book describes key principles underlying wildlife damage and its control, and uses worked examples to demonstrate the application of these principles to real-life topics.

As the book focuses on a series of principles that apply across species, it complements the species-based approach of Bruggers and Elliott (1989) to quelea finches (*Quelea quelea*) in Africa, Dobbie *et al.* (1993) to feral horses (*Equus caballus*), Caughley *et al.* (1998) to rodents in Australia and Montague (2000) to brushtail possums in New Zealand. More particularly, the book focuses on controlling the damage rather than on controlling the animal itself. Other books offer a less general discussion: Dolbeer *et al.* (1994) and VerCauteren *et al.* (2005) reviewed control of damage by North American pests, Singleton *et al.* (1999a) reviewed control of rodents, and Fenner and Fantini (1999) reviewed biological control of vertebrate pests, with emphasis on European rabbits (*Oryctolagus cuniculus*). Conover (2002) reviewed practical aspects of wildlife damage management.

The emphasis here is on good examples rather than on critiquing past studies. Hone (1994a) offers a good critical review of the subject. However, Hone (1994a) and others (e.g. Braysher 1993; Olsen 1998) do not discuss the topic of ecological theory in detail. *Wildlife Damage Control* demonstrates ecological theories such as disturbance theory, foraging theory and the theory of epidemiology and shows how they are relevant to biodiversity conservation and other topics, and how they can be evaluated in studies of wildlife damage control.

While ecology is fundamental to the book, we recognise that economic, social and political issues are also involved in management of pest damage. The need for links between disciplines and the need to put ecology into a broader perspective has been discussed by Begon *et al.* (1996, pp. 664–665), and other works have described the general human dimensions of wildlife management (Kessler *et al.* 1998; Miller & Jones 2005, 2006) and of wildlife damage control (Conover 2002). Such issues are not covered here.

Principles

Why bother with principles? Simply because principles can assist managers, scientists and others who deal with wildlife damage problems. Rather than begin each study or control program from scratch, we can learn from earlier results by identifying generalisations across different species, different locations, different methods and different times.

Reducing wildlife damage can be thought of as controlling species x at location y using method z at time t; for example, controlling rats (x) in a large city (y) by trapping (z) in winter (t). Given the wide range of wildlife damage problems, there is an almost unbelievably large number of $x/y/z/t$ combinations to consider when assessing damage and the effectiveness of control. Examples of the many options in planning control of one species in any location using one or more methods are listed in Table 1.1.

Principles also provide guidelines in the absence of other data. When a new pest emerges in a given location or when there is little available literature, generalisations from elsewhere may be useful. Principles are the building blocks upon which new data are placed. Occasionally, of course, principles have to be changed if new data show they are wrong or inappropriate in some situations. A major change in thought or approach is a 'paradigm shift' (Kuhn 1970). Such a change has occurred in this field with the shift from wildlife control to wildlife *damage* control. The new approach is described by Braysher (1993), Parkes (1993), Dolbeer *et al.* (1994), Hone (1994a), Olsen (1998) and Conover (2002).

Other fields of science, such as ecology, conservation biology and genetics, make great efforts to identify and assess principles and hence develop strategic management plans. A set of 32 ecological principles was described and discussed by Walker and Norton (1982); many were directly related to pest control and are discussed later in this book.

Table 1.1: Examples of options in the control of wildlife damage

Number	Option
Damage	
1	Control damage rather than pests
2	Target individual pests causing most damage vs no targeting
Timing of control	
3	Conduct control prior to or during damage
4	Control individuals just after their birth or at random across ages
5	Average age of pests controlled is less than the generation interval of pests
6	Control when pest population size is low or high
7	Control an increasing or decreasing pest population
8	For pest species that have a mean litter size of one, time control after or before the first breeding
Spatial aspects of control	
9	Control at sites where the economic cost of damage is greater than the cost of control
10	Control at sites where species being conserved can maintain a source population
11	Control at sites where benefits are greatest
Other options	
12	Lethal vs non-lethal population control
13	Manipulate a pest's habitat or directly change the pest's demographic rates
14	Control immigrants and emigrants rather than residents
15	Biological or non-biological control
16	Direct control to age classes contributing most young to the next generation (highest $l_x m_x$)
17	Target individuals most likely to be affected by control vs least likely to be affected
18	Involve all stakeholders or only those most likely to cooperate

Bailey (1984) stated many principles of wildlife management, with particular emphasis on biological, rather than sociological, principles. The principles were described as ‘widely accepted generalisations based on abundant and diverse research and experience and having wide application for managing wildlife’. Five principles of population dynamics were described by Berryman (1999) and are explored later in this book. Textbooks aid the development of the respective science directly, when read and used by scientists and managers, and indirectly when used as textbooks by university students. We hope that this book will make a valuable contribution to educating stakeholders, which is fundamental to successful control of pest damage (Braysher 1993; Olsen 1998).

Surprisingly, there have been relatively few attempts in the scientific discipline of wildlife damage control to describe general principles and their clear application. Braysher (1993) and Olsen (1998) suggested eight principles of managing vertebrate pests. Their principles describe general topics in science, economics, law and management. The present book aims to complement and extend those works. It will generalise the principles beyond the pests and issues in Australia, focusing on the scientific basis of pest damage control worldwide and describing how principles can and should be evaluated. Some of the principles will be familiar to land managers, pest control operators and scientists, either implicitly (unstated) or stated explicitly (rarely). An example of an implicit principle is the often-stated need for neighbours to cooperate in pest control – it is implied, though not directly expressed, that this is because of the need to limit or stop movement of pests from one property to adjacent properties. One reason for clearly stating principles is to draw out such implicit assumptions and make them explicit for all to see, discuss and evaluate.

The Concise Oxford Dictionary (Sykes 1976) defines a principle as, among other things, a 'general law as guide for action'. Although principles can be powerful tools they should not be considered beyond evaluation. We should not simply accept them as general laws. Their relevance to a particular damage type, pest, location, time and control method can be assessed if needed. General principles can be evaluated as hypotheses in the sense of the scientific method illustrated by Krebs (2001, Fig. 1.3). The development and evaluation of hypotheses will generate more reliable knowledge and stronger inferences (Platt 1964) in wildlife ecology and management (Romesburg 1981; Macnab 1983).

Principles may become more complex as our understanding of nature improves. For example, a principle could state that after pest control has reduced a population, the abundance of herbivorous pests increases exponentially at the intrinsic rate of increase (r_m) (see Chapter 3 for further discussion). The exponential increase is known as Berryman's (1999) first principle of population dynamics. Such an increase may be correct in many situations, but if there is a significant predator of the pest at the site its predation may change the pattern of population increase. The rate of increase may be slower than r_m or there may be no increase at all, depending on site conditions such as climate and food. These variations are explored in Berryman's (1999) second to fifth principles of population dynamics. The original principle changes and may become more complex.

Some works do not clearly state the principles they are discussing. Readers are expected to work out which points or sentences are the principles, and may get it wrong. Each chapter in this book includes a summary table of principles. Examples from the literature are given wherever possible; if there are no good examples then hypothetical examples are given. Readers are encouraged to evaluate the examples and principles, especially as good data become available.

Two types of principles are used here. First are principles that are generalisations, derived by induction, from studies across many species, locations, methods and times. Second are principles that are hypotheses, proposed for further evaluation by deduction. These are clearly described as hypotheses.

Framework

A guiding framework is helpful in assessing the significance of damage or impacts by vertebrate pests. The framework is represented in a series of diagrams, introduced and explained below. It encourages analysing the relationships between pest density and other response variables, and between the level of pest control and other response variables. Broadly speaking, it questions the relationships between the levels of inputs and the levels of outputs.

A traditional approach in the study of wildlife control is to first study pest dynamics then infer or estimate the effects of various types of pest control on pest dynamics (Fig. 1.1a). Sometimes it extends to include the dynamics of the pest relative to its food supply and its predators in a trophic structure (Fig. 1.1b). While this approach tells us about the effect of one level of a control (e.g. poisoning) on pest density, it does not tell us about the effects of other levels of control. What happens, for example, if we use only half the amount of poison? What happens to production yield with different levels of pest control – does using more poison always result in higher production?

An alternative paradigm, used in this book, studies crop yield (or some other response variable) relative to both pest dynamics and the effects of pest control (Fig. 1.2). The emphasis has shifted from estimating the

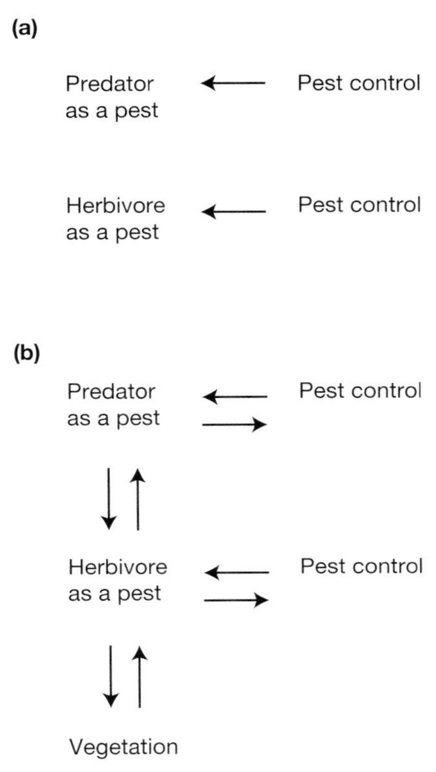

Figure 1.1: A traditional research paradigm describing the effects of pest control on pest population dynamics. (a) Historical view, which focused on the effects of pest control on pests only. (b) Trophic structure view, which recognised the effects of pest control on one or all of predators, herbivores and vegetation.

effect of a treatment (e.g. poison vs no poison) on a pest's dynamics to focus on the pest's effects, such as estimating the relationship between a response (e.g. yield) and an input (e.g. amount of poison). Such a change has already been encouraged elsewhere in wildlife research (e.g. Steury *et al.* 2002). The ideas can be applied readily to wildlife damage.

The new broader approach can also be shown as a graph relating the effects of a pest to the pest's density (see Fig. 1.3). Figure 1.4 shows diagrammatically the effects on a response variable (e.g. crop yield) of different levels of pest control effort (e.g. different numbers of traps). The focus is on the *relationship* between pest density and yield (or other response variable) and costs of control (Fig. 1.3), and the *relationship* between the level of pest control efforts and yield (or other response variable) and costs of control (Fig. 1.4). The principles for managing wildlife damage discussed in this book often describe relationships between variables. As

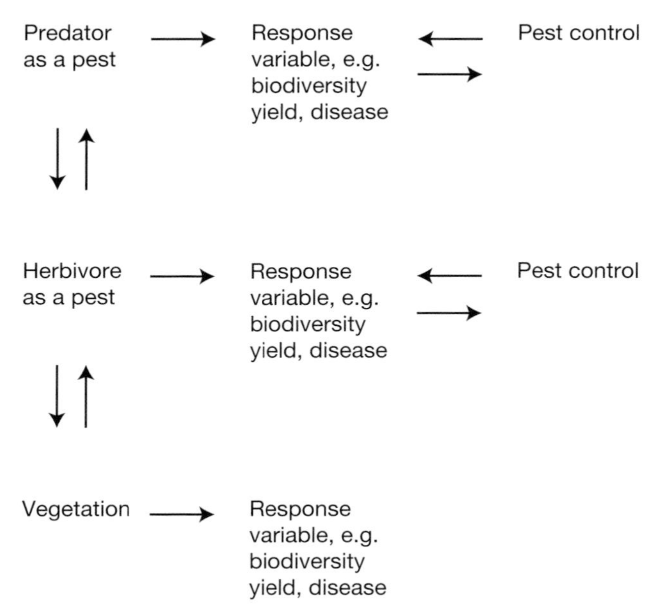

Figure 1.2: An alternative research paradigm, describing the effects of pest control on pest damage or other response variable, and the effects of pest population dynamics on the same response variable.

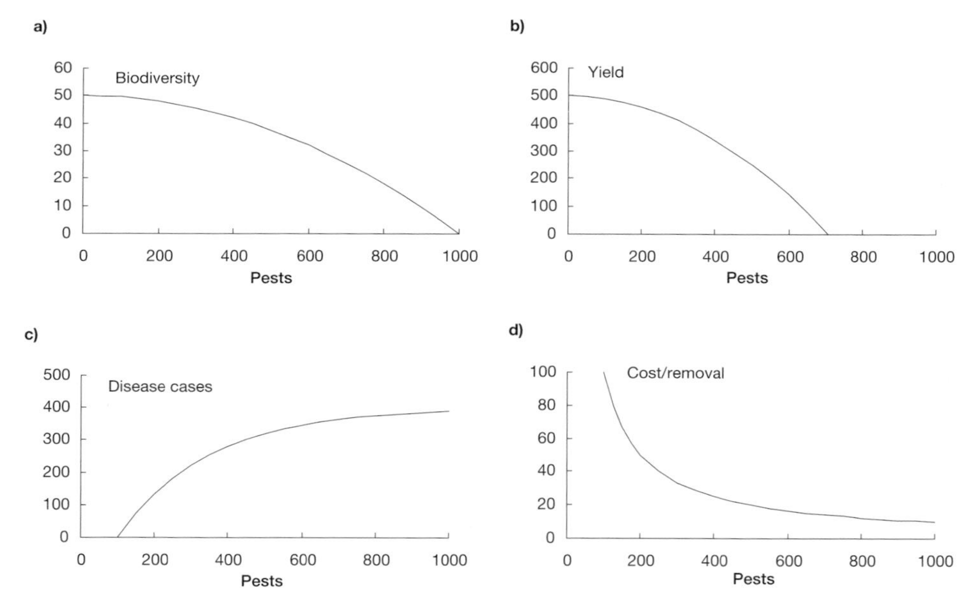

Figure 1.3: Analysis of general relationships between pest density and response variables is important. Response variables include (a) biodiversity, (b) yield of a crop, livestock activity, forest timber or fishery, (c) incidence of disease, and (d) the per capita cost of pest control during the phase of reducing pest density. In each graph, the x axis is pest abundance or population density. Note: the specific form of any relationship may not be exactly as shown.

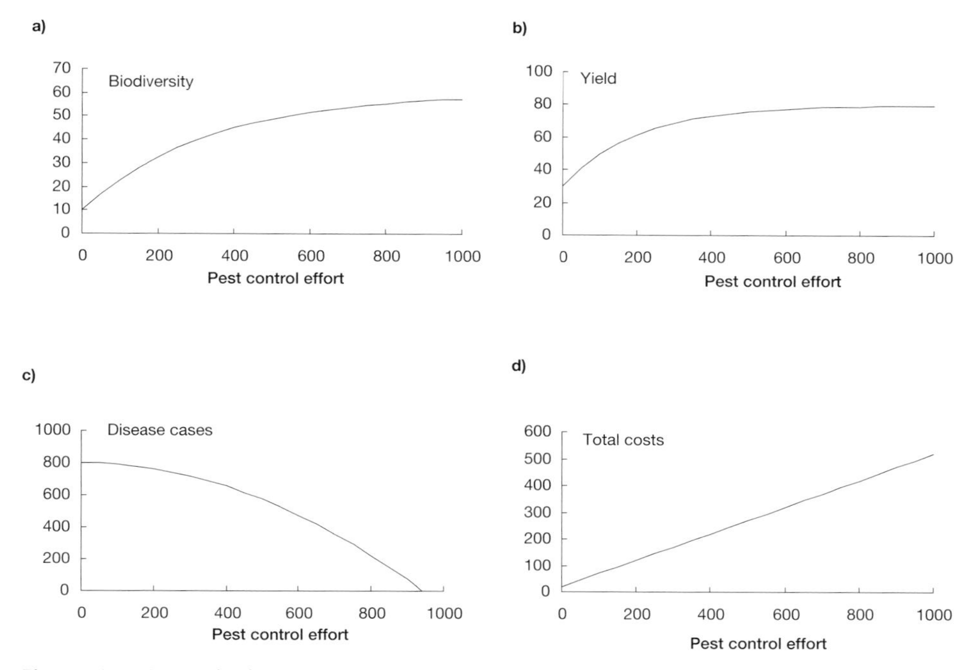

Figure 1.4: General relationships between level of pest control effort and response variables. Response variables include (a) biodiversity, (b) crop yield, (c) incidence of disease, and (d) the total cost of pest control. In each graph, the x axis is pest control effort. Note: lines or curves are examples of general relationships and the specific form of any relationship may not be exactly as shown. The scale is arbitrary.

such, they can be shown in a graph and the values of parameters in the relationships can be estimated, for example by regression analysis.

In summary, the ideas represented in Figures 1.2 to 1.4 represent a framework for analysing wildlife damage control. The specific details of each pest, each location, each control method and each time period fit into the framework, which will be frequently referred to throughout the book.

The ideas shown in Figure 1.4 can be inputs into simple economic analyses, such as cost–benefit analysis. This is particularly useful in activities such as agriculture and forestry. In other fields, such as conservation of biodiversity, people may use value judgements with no specific economic basis – they simply want to save a species or community from local or global extinction. Similarly, economics may not always be the basis of an evaluation when assessing the value of control for human health.

Several issues need to be considered when applying the above framework.

First, it is not pest abundance that by itself causes problems. It is the reduction in yield, for example, caused by pests that makes us consider them 'pests'. Accordingly, the aim of wildlife damage control should be to reduce the damage rather than control the pest itself. Braysher (1993) identified damage control as an important part of managing vertebrate pests. The relationship between pest abundance and pest damage is discussed in Chapter 2.

Second, if the aim of wildlife damage control is damage reduction it is necessary to demonstrate – not assume – that damage is being caused by a particular species. Damage may be caused by a different species or be a by-product of poor management. There may sometimes be uncertainty about what is causing the damage. This occurred in southern Australia where some farm managers assumed that lamb loss was caused by predation by feral pigs (*Sus scrofa*), while other farmers thought low lamb production was due to lambs getting lost in long grass and dying of starvation. Research by observation (Pavlov & Hone 1982) and field experiment (Pavlov *et al.* 1981; Choquenot *et al.* 1997) showed that the feral pigs were a significant predator of lambs.

Third, the decision to undertake damage control may be made without firm data. Many such decisions occur daily. For example, a farmer may have to decide to do control of house mouse before it is certain that a mouse plague (and hence considerable damage to crops and stored grain) will occur. If the farmer waits until a mouse plague has started before commencing control it will likely be too late; there will be so many mice that control efforts cannot reduce or stop damage. The control may be initiated because the potential consequences of not doing control are unacceptable, even though damage may be unlikely to occur.

Fourth, a mix of methods can be used in wildlife damage control studies, as is done in ecology (Krebs 2001). Typically, many studies of pest damage control begin with simple observations. White (1775) described an observational approach when studying potential agricultural damage by rooks (*Corvus frugilegus*) in southern England: 'A rook should be shot weekly the year through and its crop examined; hence perhaps might be discovered whether on the whole they do more harm than good'. More recently, observations of bovine tuberculosis in wildlife, such as badgers in the UK and brushtail possums in New Zealand, resulted in the control of bovine TB changing from a focus on disease control in cattle alone to include actions against wildlife. After such initial

Table 1.2: A classification of scientific studies from classical experiments to simple observations, based on the presence or absence of treatment randomisation, treatment replication, simultaneous experimental controls and statistical analysis

	Experimental control used	Treatment randomisation used	Treatment replication used	Analysis conducted
Classical experiment	✓	✓	✓	✓
Data set awaiting analysis	✓	✓	✓	
Unreplicated experiment	✓	✓		✓
Unreplicated, unanalysed experiment	✓	✓		
Quasi-experiment type I	✓		✓	✓
Quasi-experiment type II	✓		✓	
Quasi-experiment type III	✓			✓
Quasi-experiment type IV	✓			
Pseudo-experiment type I		✓	✓	✓
Pseudo-experiment type II		✓	✓	
Pseudo-experiment type III		✓		✓
Pseudo-experiment type IV		✓		
Pseudo-experiment type V			✓	✓
Pseudo-experiment type VI			✓	
Pseudo-experiment type VII				✓
Simple observations				

observations, experiments are often carried out to clarify the effects of particular actions (treatments). If experiments are not possible then modelling may be used to predict effects of the actions, provided the modelling is based on real data and realistic assumptions.

Studies can range from observations to classical experiments: a variety of possible study types between these two extremes contain elements of both observation and classical experiment. Table 1.2 lists some studies

Table 1.3: Characteristics of published field studies of vertebrate pest control (n = 43), control of undesirable diseases (n = 21) and evaluations of biological control of vertebrate pests (n = 22)

	Vertebrate pest control	Control of undesirable diseases	Biological control
Classical experiment	8	0	0
Data set awaiting analysis	0	0	0
Unreplicated experiment	0	0	0
Unreplicated, unanalysed experiment	0	0	0
Quasi-experiment type I	13	1	0
Quasi-experiment type II	2	3	2
Quasi-experiment type III	3	1	0
Quasi-experiment type IV	6	2	1
Pseudo-experiment type I	2	0	0
Pseudo-experiment type II	0	0	0
Pseudo-experiment type III	0	0	0
Pseudo-experiment type IV	0	0	0
Pseudo-experiment type V	2	1	1
Pseudo-experiment type VI	2	7	9
Pseudo-experiment type VII	2	0	1
Simple observations	3	6	8

that do, or do not, have essential features of experiments. There are many types of classical experiments (Manly 1992; Underwood 1997; Krebs 1999) and many types of more elementary designs, such as pseudo-experiments and quasi-experiments, that fall between classical experiments and observations (Table 1.2). A useful discussion of those elementary designs is given by Manly (1992, Ch. 3) and Manly (2001, Ch. 4). There is ongoing debate about the strengths and weaknesses of the various methods (Krebs 1988; McCallum 1995), but a mix of methods is still the most common practice.

Systematic collation of data on the characteristics of published studies shows that there is a need to distinguish between the many types of study designs. Table 1.3 shows studies of vertebrate pest control that aimed to estimate the effect of a pest control method were mostly not classical experimental designs. Similarly, studies of control of undesirable diseases and evaluations of biological control of vertebrate pests were often simple observations. In Table 1.3, the pest control studies were examples of manipulative experiments examining the effects of treatments, such as poisoning and shooting, on pest abundance. Studies have been classified according to whether they reported using a simultaneous experimental control, randomisation, replication and statistical analysis, using the format of Table 1.2. Where studies use more observational methods the results should be interpreted as such, and not as equivalent to the results of classical experiments (McArdle 1996). Ecological studies often lack key design features and can have limited statistical power (Raffaelli & Moller 2000), like many of the studies in Table 1.3.

Because of the variety of study designs, published studies of wildlife damage control can reach a range of conclusions or make a range of inferences. Studies of a more observational nature can make weak inferences about cause and effect (*maybe* predator control reduced livestock predation) and studies that involve classical experiments can make stronger inferences (predator control *did* reduce livestock predation) (Platt 1964; McArdle 1996). Table 1.4 shows the varying strength of inferences that can be obtained from studies of predation control. The study design characteristics (from Table 1.2) are also shown. Simple observations yield direct observation of predators killing livestock, which allows the scientist or manager to say little about the effects of predation control (weak inference). Classical experiments directly estimate the effects of predation control in randomised, replicated studies which allow the scientist or manager to reach strong conclusions (make stronger inferences).

Wildlife damage control has a long history. As pest problems rarely go away or are solved quickly, operators and managers need to learn what past and current efforts have achieved and how to do better to fulfil the aims of pest control. Adaptive management is a method of learning and developing better damage control. It involves deciding what to do, doing it, then assessing what happened (usually by monitoring), modifying actions next time if the aims were not achieved, repeating the assessment and so on (Walters 1986; Braysher 1993; Lancia *et al.* 1996; Walker 1998).

Conclusion

In summary, this book sets out a series of principles for controlling wildlife damage and discusses how the principles have been used in management actions in the past or could be used in future. The principles described throughout the book should not be taken as dogma: readers should ask

Table 1.4: Increasing strength of inference that management action of predation control caused a change in livestock production. The action is assumed to be lethal predator control

Design and data	Strength of inference
Simple observations	
Predators known previously to kill livestock	*Weak*
Control known previously to kill predators	
Alternative explanations not evaluated	↓
Predation inferred from signs, e.g. carcasses	
Predation observed once	↓
Predation observed many times	
Pseudo-experiment	↓
No experimental controls used	
Quasi-experiment	↓
A different response occurs in experimental controls and treatment areas	
Predators killed by control	↓
Predator abundance reduced by control	
Predation reduced by control	↓
Positive relationship observed between predation (damage) and predator abundance	
Positive relationship observed between production (yield) and control effort	↓
Classical experiment	
Production is higher in randomised treatment areas than in randomised experimental control areas	↓
Alternative explanations evaluated and rejected	
All of the above (except first three)	*Strong*

questions such as how they can estimate the key relationships described in the principles, the consequences of not using those principles, and what they are forcing themselves or others to do if principles are not applied.

Chapter 2 examines generalities in patterns and processes of wildlife damage and Chapter 3 looks at generalities in controlling wildlife damage. Readers will benefit from reading Chapters 2 and 3 before reading later chapters on particular damage topics.

2

Patterns and processes in wildlife damage

Introduction

The types and extent of damage caused by wildlife are many and varied. However, knowledge of the extent of such damage, what determines the extent of damage and why it varies, may be limited. As stated in Chapter 1, if managers and scientists had to determine the types and extent of damage for each situation there would be a very large number of possible results. But life need not be so difficult.

Consider predation of livestock. In different parts of the world, wildlife are important predators of livestock. In North America, coyotes kill sheep (Conner *et al.* 1998; Sacks *et al.* 1999). Wolves kill livestock in Europe (Meriggi & Lovari 1996), wolverine (*Gulo gulo*) kill sheep in Norway (Landa *et al.* 1999) and lynx kill sheep in France (Stahl *et al.* 2001a, b, 2002). Tibetan wolves (*Canis lupus chanku*) and snow leopards (*Uncia uncia*) reportedly prey on livestock in elevated parts of India (Mishra 1997). Red foxes (Rowley 1970; Greentree *et al.* 2000) and feral pigs (Pavlov & Hone 1982; Choquenot *et al.* 1997) kill lambs in Australia. What do these examples have in common?

The examples demonstrate common patterns and processes such as relationships between kills and prey abundance, between kills and predator abundance and the behaviour of predators. These examples of livestock predation have more than a passing connection to examples of

Table 2.1: Principles relating to the level and variation in damage by vertebrate pests

Determinants of damage
2.1: Damage extent. The extent of pest damage is related to pest abundance, prey (or the availability of what is being damaged), and landscape features
2.2: Damage response determinants. Response variables (e.g. yield) are related to the level of pest damage and pest density
Variation in damage
2.3: Individual heterogeneity. Heterogeneity exists between individuals of a species in the extent of damage they cause
2.4: Damage variation. Damage varies spatially and temporally

predation in population ecology, as described by Begon *et al.* (1996) and Krebs (2001). The components of predation damage can be generalised to other wildlife damage topics, such as crop and forest damage, as a set of principles.

This chapter describes four general principles relating to the patterns and processes of damage caused by vertebrate pests that can apply across a wide range of pest problems. Specific aspects of damage to production, conservation, health and recreation are discussed in detail in later chapters. The chapter links the principles to the framework relationships depicted in Chapter 1 (Figs 1.2–1.4). It discusses two types of principles: those about determinants of damage (which focus on core relationships), and those relating to variation in damage (which focus on variation around those relationships). The principles are listed in Table 2.1.

Determinants of damage

The following principles regarding determinants of damage focus on core relationships such as those between yield and damage and between damage and pest abundance.

Damage, pests and landscape

In wildlife damage control work it is often assumed that there is a positive relationship between the number of pests and the damage they cause. This is borne out by most empirical studies, which show that the extent of pest damage is related to one or more variables such as the abundance of pests, their prey (or the availability of what is being damaged) and landscape features. Examples of the relationship between pest damage and pest abundance have been seen in many studies, including a study of damage to oats and barley crops by rooks in Scotland (Feare 1974) and predation of lambs by feral pigs in New South Wales (Fig. 2.1). A collation of empirical data from 39 such studies reported that 21 (54%) showed a significant linear relationship (Hone 1996). A more recent collation showed the range of damage topics and evidence for the relationship between damage and abundance (Table 2.2). The table splits data on pests into those causing environmental damage, production damage, and damage based on diseases and health issues. The linear or curved (concave up or down) nature of the relationship is also given. Linear relationships are positive unless shown otherwise. Note that tests of shape invariably do not force the regression through the origin.

However, mixed results have also been reported. One such example was in a study on predation of sheep by wolverine in Norway. A step-wise multiple regression found a significant positive relationship between sheep losses and sheep abundance, but no effect of wolverine abundance (Landa *et al.* 1999). However, the percentage loss of ewes was positively related to wolverine abundance.

The form of the relationship between damage and pest abundance has also been explored theoretically, as a curved relationship (Headley 1972; Tisdell 1982; Izac & O'Brien 1991) or a linear and curved relationship (Barlow 1987; Braysher 1993; Hone 1994a; Bomford *et al.* 1995; Conover 2002; Hone 2004). Sigmoidal relationships between damage to crops

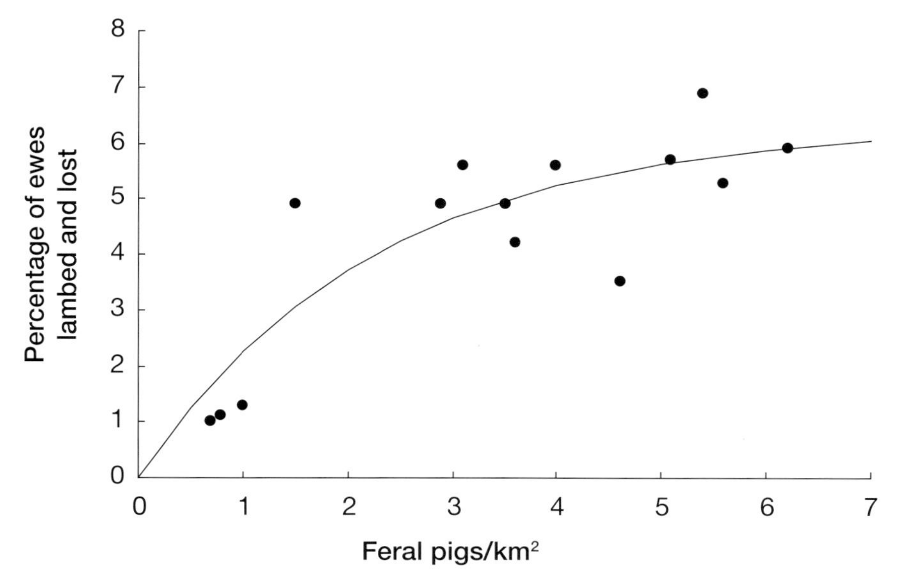

Figure 2.1: The relationship between an index of lambs killed by feral pigs and feral pig population density in western New South Wales.

Source: After Choquenot *et al.* (1997)

Table 2.2: Testing the relationship between pest damage (y) and pest abundance (x)

Damage	Pest	Relationship	Significance	Source
Damage to environment				
Ground rooting	Feral pig	Linear	$P < 0.05$	After Ralph and Maxwell (1984)
Changes in ground rooting	Feral pig	Linear	$P < 0.05$	Hone (1988a)
Ground rooting	Feral pig	Linear	$P < 0.01$	Hone (1988b)
Grass nodes	Wallabies	Linear	$P < 0.01$	Ramsey and Engeman (1994)
Grass stolons	Wallabies	Linear	$P < 0.01$	Ramsey and Engeman (1994)
Change in ground rooting	Feral pig	Curve; down	$P < 0.05$	Hone (1995)
Plant foliage cover	Brushtail possum	Linear	$P < 0.05$	Pekelharing *et al.* (1998)
Water vole	Mink	Linear	$P < 0.05$	Barreto *et al.* (1998)
Coot abundance	Mink	Linear	$P < 0.05$	Ferreras and Macdonald (1999)
Ground rooting	Feral pig	Curve; down	$P < 0.01$	Hone (2002)
Ground rooting	Feral pig	None	NS	After Cooray and Mueller-Dombois (1981)
Ground rooting	Feral pig	None	NS	Hone (1988a)
Moorhen abundance	Mink	None	NS	Ferreras and Macdonald (1999)
Damage to products				
Oats	Rook	Curve; down	$P < 0.01$	Feare (1974)
Barley	Rook	Curve; down	$P < 0.05$	Feare (1974)
Crops	Wild boar	Linear	$P < 0.01$	Gorynska (1981)
Wheat	Lesser bandicoot rat	Linear	$P < 0.01$	Poche *et al.* (1982)
Corn	Red-winged blackbird	Linear	$P < 0.01$	Bollinger and Caslick (1984)
Pistachio nuts	Crow	Linear	$P < 0.05$	Crabb *et al.* (1986)
Pistachio nuts	Scrub jay	Linear	$P < 0.05$	Crabb *et al.* (1986)
Sugarcane	Rat	Linear	$P < 0.01$	Lefebvre *et al.* (1989)
Wheat	Brent goose	Linear	$P < 0.05$	Summers (1990)
Soyabean	House mouse	Linear	$P < 0.05$	After Twigg *et al.* (1991) and Singleton *et al.* (1991)

Damage	Pest	Relationship	Significance	Source
Wheat	House mouse	Linear	$P < 0.01$	Mutze (1993)
Wheat	House mouse	Linear	$P < 0.01$	Mutze (1993)
Pasture biomass	Red-necked wallaby	Linear	$P < 0.05$	After Statham (1994)
Lamb predation	Feral pig	Curve; down	$P < 0.05$	Choquenot *et al.* (1997)
Scots pine bark-stripping	Red squirrel	Linear	$P < 0.05$	Bryce *et al.* (1997)
Ewe predation	Wolverine	Linear	$P < 0.005$	Landa *et al.* (1999)
Crop	African elephant	Linear	$P < 0.005$	Hoare (1999)
Grass	Plateau zokor	Linear	$P < 0.01$	Fan *et al.* (1999)
Peanuts	Rodent	Linear	$P < 0.001$	Zhang *et al.* (1999)
Wheat	Zokor	Linear	$P < 0.01$	Zhang *et al.* (1999)
Corn	Zokor	Linear	$P < 0.01$	Zhang *et al.* (1999)
Rice	Vole and mouse	Linear	$P < 0.01$	Zhang *et al.* (1999)
Loss in pasture height	Rabbit	Linear	$P < 0.001$	Fleming *et al.* (2002)
Crop	Wild boar	None	NS	Mackin (1970)
Crop	Wild boar	None	NS	Andrzejewski and Jezierski (1978)
Barley	Duck	None	NS	Gillespie (1985)
Pine tree	Rodent	None	NS	Murua and Rodriguez (1989)
Oilseed rape	Brent goose	None	NS	McKay *et al.* (1993)
Damage via diseases				
Prevalence of bovine TB	Badger	Linear	$P < 0.01$	Hone (1994a); after Cheeseman *et al.* (1981)
Prevalence of bovine TB	Badger	Linear (negative)	$P < 0.05$	After Cheeseman *et al.* (1989)
Prevalence of bovine TB	Brushtail possum	Linear	$P < 0.05$	After Coleman (1988)
Prevalence of TB in ferrets	Brushtail possum	Curve (positive)	$P < 0.001$	Caley *et al.* (2001)
Prevalence of TB in ferrets	Ferrets	Curve (negative)	$P < 0.001$	Caley *et al.* (2001)
Prevalence of bovine TB	Badger	None	NS	After Cheeseman *et al.* (1989)

Damage	Pest	Relationship	Significance	Source
Damage via health and accidents				
Aircraft strikes	Bird	Linear	$P < 0.01$	After van Tets (1969)
Deer road kills	Deer (buck harvest)	Linear	$P < 0.05$	McCaffery (1973) (28 of 29 tests)
Deer road kills	Deer (buck harvest)	None	NS	McCaffery (1973) (1 of 29 tests)
Aircraft strikes	Gulls	None	NS	Burger (1985)

Source: Modified from Hone (1994a, Table 2.1).
Note: A source cited as 'After ...' indicates that analysis was based on data from that source but was not in the original study.

and tree seedlings and deer (*Odocoileus* spp.) abundance were considered likely by Conover (1997b, 2002). The different relationships are alternative, or multiple, hypotheses in the sense of Chamberlin (1965).

From such empirical and theoretical studies we can generalise a first principle of wildlife damage, **damage extent** – that the extent of pest damage is related to pest abundance, prey (or the availability of what is being damaged) and landscape features (Table 2.1). The damage extent principle is the most fundamental principle relating to pest damage.

The above discussion assumed that when there were no pests there was no damage – the relationship between pest damage and pest density went through the origin ((0,0), see Fig. 2.1). An alternative hypothesis is that the relationship starts not at the origin but to the right of the origin at some threshold pest density, hence the line cuts the x axis (e.g. at (17,0), Fig. 2.2a). The presence of a threshold has important implications for pest control – it means that not all pests have to be removed in order to stop damage caused by the pests. Instances of a threshold pest density are common with hosts of undesirable infectious diseases (discussed further in Chapter 6) where the pest density has to exceed the threshold number for the disease to spread into a population of susceptible hosts (Anderson & May 1979). An example of a threshold population density relates to foxes and rabies incidence in Europe (Anderson *et al.* 1981). Similarly, a theoretical relationship between a measure of damage, mean parasite level of an undesirable worm, and equilibrium host (pest) density is curved (Fig. 2.2b) (Dobson & May 1991).

Alternatively, a threshold may occur on the y axis (Fig. 2.2c). The intercept may be above the origin, where there are zero pests but damage is estimated to be greater than zero. Likely reasons for such a threshold

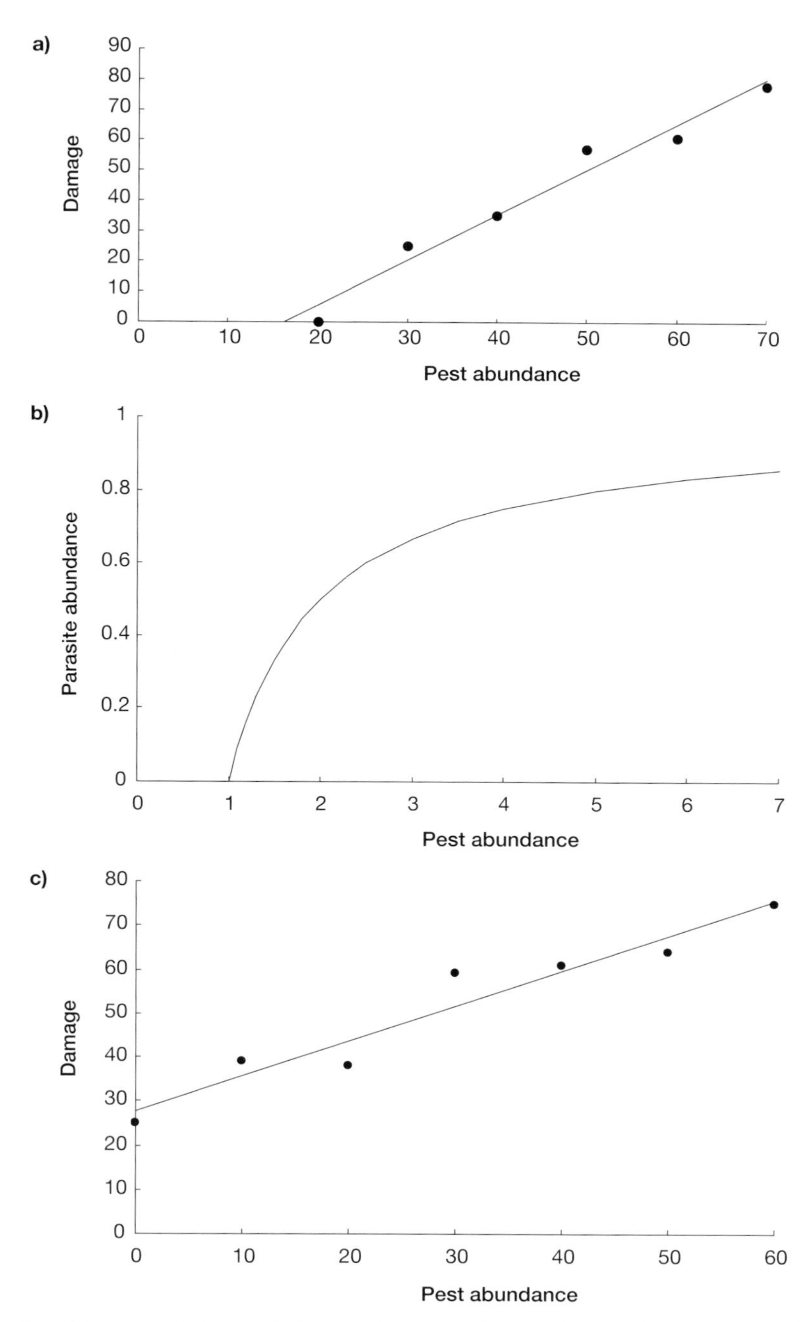

Figure 2.2: (a) A hypothetical relationship between damage by vertebrate pests and abundance of the pests, with a threshold level of pest density. (b) The theoretical equilibrium relationship between mean density of a directly transmitted infectious parasite and abundance of the hosts (after Dobson & May 1991). (c) A hypothetical threshold that would be caused by bias in methods of estimating damage, or damage being caused by a different pest species.

are bias in the damage estimates or that the damage is actually being caused by pests other than the species being studied (Fig. 2.2c). To test for the presence of a threshold on the x or y axis we use a one-tailed (not two-tailed) test.

The relationship between damage, pest density and other determinants can be estimated by regression analysis from observational and experimental studies. Useful textbooks on regression include Manly (1992) and Underwood (1997). Such estimating uses a mix of methods and experimental design principles. In observational studies, ideally, data are collated from different sites with different numbers of pests and levels of damage. The advantage of this method is that levels of pest density do not have to be predetermined. The main disadvantage is that pest density may vary over a limited range of population densities. In observational studies most sites commonly have little or no damage and a few sites have a lot of damage, which can make it difficult to estimate the relationship between levels of pest density and damage.

An alternative approach is to establish an experiment with different levels of pest density. This approach was used to estimate the relationship for various measures of agricultural production and abundance of European rabbits (Croft 1990; Croft *et al.* 2002; Fleming *et al.* 2002; Hone 2006). The experimental approach has the advantage that pest numbers can be set at a level considered appropriate based on experience, precedents (results in the literature) and relevant theory (linear vs curved effects).

In both observational and experimental studies there is always a chance of non-significant results, which corresponds to evidence of no relationship between pest damage and pest density. Before accepting a non-significant response, we should consider likely reasons for the result (Table 2.3). First, the variables may be measured imprecisely (with a high variance) so a type II error has occurred (i.e. the relationship is real but has not been detected). Second, an *a priori* model is incorrectly specified, for example when the underlying relationship is a curve not a straight line (again a type II error has occurred). Third, what has been measured as damage is not linearly related to actual damage. Fourth, other sources of variation were not included in the analysis so the strength of any underlying relationship between damage and pests has been underestimated (a type II error has occurred). Fifth, data may come from the 'flat' part of a curve so a curve was not detected. A sixth

Table 2.3: Likely reasons for a non-significant result in a study designed to estimate a relationship between two variables

Number	Reason
1	Variables measured imprecisely
2	Model specified incorrectly
3	Response variable measured incorrectly
4	Other sources of variation in the response variable not measured
5	Range of variation in the independent variable was too narrow
6	Non-significant result is real

possibility is that a non-significant result is real. This could occur if a few individual pests do the damage and their effects are not correlated with pest abundance. In the absence of additional data these possibilities cannot be distinguished easily. However, they may be differentiated during a pilot study, and the likelihood of a type II error reduced by designing the subsequent study using the results of a power analysis of data from the pilot study.

In population ecology the relationship between kills and prey abundance is described in the per capita functional response. Many functional responses have been demonstrated in ecology (Begon *et al.* 1996; Krebs 2001). In wildlife damage control, the analogous relationship is between how much damage has been done and how much is available to be damaged. The extent of pest damage to a crop, in absolute terms, is assumed to be related positively to how much crop is available to be damaged: the more crop, the more damage. An example directly relating to wildlife control is the positive relationship that has been observed between the number of lambs killed by feral pigs per week and the abundance of the lambs (Hone 1994a, Fig. 2.5). Other examples of functional response relationships in wildlife control involve red kangaroo (*Macropus rufus*), European rabbit (Short 1985), red fox (Pech *et al.* 1992), feral pig (Choquenot 1998) and house mouse (*Mus musculus*) (Ruscoe *et al.* 2005). When estimating a functional response, sample sizes need to be large to confidently discriminate type II and type III functional responses (Marshal & Boutin 1999).

Yield and damage

Simple logic suggests that as the extent of damage increases the yield of agricultural products such as crops will decline. A number of studies bear this out. Rooks reduced plant density in oat and barley crops in parts of Scotland, causing reduced yield (Feare 1974). As a result there was a positive curved relationship between rook abundance and loss in crop yield (see Fig. 2.3a, a form of the framework relationship shown in Figure 1.3b). Grazing by pink-footed (*Anser brachyrhynchus*) and greylag (*Anser anser*) geese was associated with a decline in yield of autumn-sown barley in Scotland in field studies (Patterson *et al.* 1989) and simulated grazing studies (Abdul Jalil & Patterson 1989). Yield of winter wheat decreased as an index of brent geese (*Branta bernicla*) abundance increased

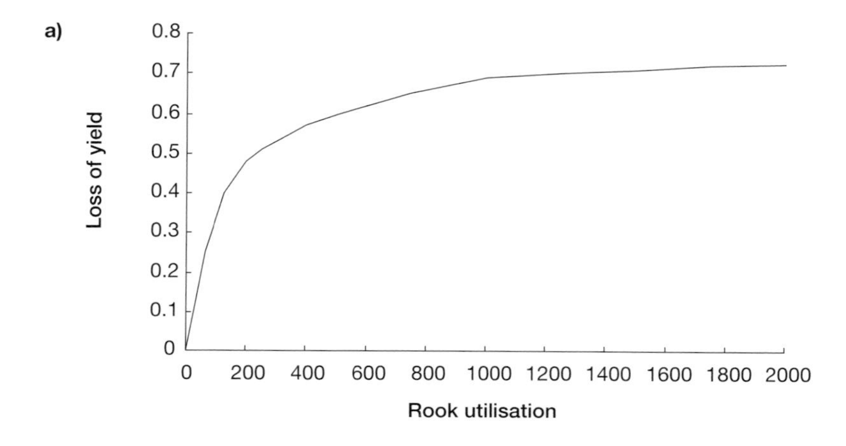

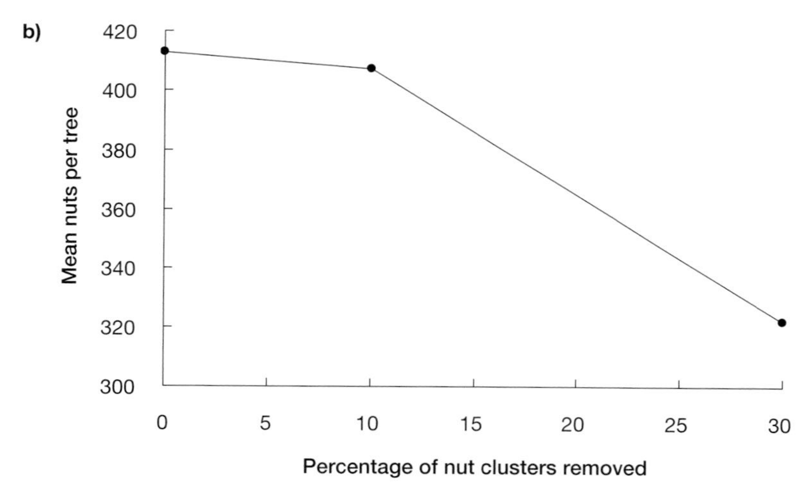

Figure 2.3: (a) The relationship between the estimated loss of yield of oats in north-east Scotland caused by rooks, and rook abundance (after Feare 1974). (b) The observed relationship between yield of macadamia nuts per tree and rat damage in Hawaii. If compensation did not occur the relationship would be linear, but analysis provided evidence of a non-linear relationship (after Tobin *et al.* 1997).

in parts of Britain (Summers 1990). An experimental study in Australia of the effects of European rabbits showed that as rabbit abundance increased, pasture height decreased and sheep liveweight decreased (Croft 1990; Croft *et al.* 2002; Fleming *et al.* 2002; Hone 2006). A decline in yield of macadamia nuts in Hawaii with increasing damage by rats (Fig. 2.3b) was demonstrated by Tobin *et al.* (1997).

The agricultural concept of 'yield' can be applied to other topics, such as biodiversity conservation. For example, a measure of species abundance or community diversity could be used as the response variable instead of yield. One such response variable is species richness in a community. This was demonstrated in south-eastern Australia, when the plant species richness of subalpine grassland was found to decrease as the extent of ground rooting by feral pigs increased (Hone 2002). Such biodiversity effects may vary with site conditions, such as nutrient-poor versus nutrient-rich soils (Proulx & Mazumder 1998), as discussed in Chapter 4.

The relationship between yield and damage may not be linear, however, as plants or animals not killed or damaged may compensate. These surviving plants and animals may have increased survivorship or increased yield because of the extra resources now available to each of them. Such compensation has been suggested to occur in wheat damaged by rodents (Poche *et al.* 1982), sunflowers damaged by birds (Cummings *et al.* 1989) and macadamia nut trees damaged by rats (Tobin *et al.* 1993, 1997). One study found that highest wool production ($/ha) occurred at intermediate levels of rabbit density (Fleming *et al.* 2002). The possibility of compensation was reviewed by Hone (2004).

In population ecology, the topic of compensation by plants was reviewed by Belsky (1986) and Ritchie and Olff (1999). The latter review considered compensation from the plant community rather than from the perspective of a single species. An analogous concept of compensation occurs for wildlife in population ecology, for individuals not killed by predators or harvesting. Such individuals have higher survivorship than that expected if mortality was additive (Nichols *et al.* 1984; Nichols 1991).

From such studies we can suggest a generalisation about yield or other response variables. This is the second principle of wildlife damage, **damage response determinants** – that response variables (e.g. yield) are related to the level of pest damage and pest density (Table 2.1). This principle has management implications because yield (or other response variable) is typically managed by varying pest density or changing the behaviour of particular individuals that cause the most damage. The latter aspect is examined below.

Variation in damage

Relationships between pests and the damage they cause are not perfect; they show variational scatter around the relationship. Principles relating to variation in damage focus on variation around the core relationships.

Pest heterogeneity

To manage wildlife damage, it is important to know whether all individuals in a population are contributing or whether the damage results from selected individuals. For example, crop damage by Asian elephants (*Elephus maximus*) in southern India was mostly caused by adult male elephants (Sukumar 1991). Predation by coyotes on sheep in California was mostly by breeding coyotes (Sacks *et al.* 1999). Predation of lambs by feral pigs in New South Wales was mostly by adult male pigs (Pavlov & Hone 1982). Attacks on people by Australian magpies are nearly always by male magpies (Cilento & Jones 1999) but involve only some individuals, as only around 10% of magpies attack (Jones 2002). There was also variation among individual lynx in killing sheep in eastern France (Stahl *et al.* 2002). In a review of many such studies of variation between individuals, Linnell *et al.* (1999) concluded that 'problem individuals' could occur, and that adult males killed disproportionately more livestock than other age and sex classes. More intensive individually based field studies were recommended (Linnell *et al.* 1999). From such studies we can draw the third principle of patterns and processes in wildlife damage, **individual heterogeneity** – that heterogeneity exists between individuals of a species in the extent of damage they cause (Table 2.1).

The concept of variation in damage between individual pests is analogous to the concept of keystone species in community ecology. Keystone species are those that affect a community far more than the abundance of that species would indicate (Krebs 2001). Obviously, however, the variation in damage concept is at the individual level of organisation while the keystone concept is at the community level. The application of the keystone species concept is explored further in Chapter 4.

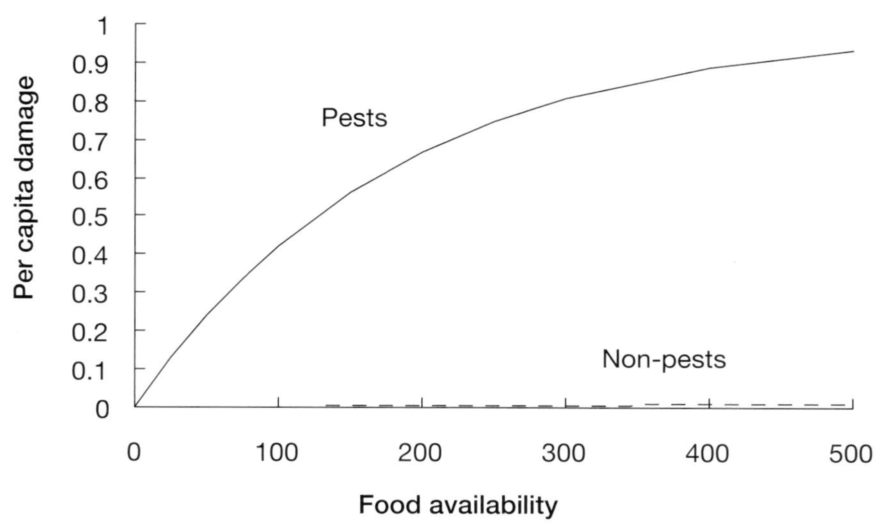

Figure 2.4: Hypothetical functional response relationships for two segments of a pest population. The solid line shows damage caused by individuals in segment 1 (pests) and the dashed line shows damage caused by individuals in segment 2 (non-pests). The dashed line is shown slightly above the x axis so that it is visible. After Hone (2004).

The concept of heterogeneity between pests, a relationship between damage and pests, and the functional response may appear contradictory but it need not be. When heterogeneity occurs the pest population may be thought of as comprising several segments, for example one segment that causes damage and a second segment that causes no damage. Individuals in each segment have different functional responses: the pests have a particular functional response and the non-pests have no functional response (Fig. 2.4). Total damage is the sum of damage from all the pests in the population.

Spatial and temporal variation in damage

Damage typically varies across different locations and at different times. Predation by coyotes of sheep and lambs varied between months in northern California (Conner *et al.* 1998). In eastern France, 33–69% of sheep kills by lynx occurred in less than 5% of the study area (Stahl *et al.* 2001a). Damage by brown bears (*Ursus americanus*) to forestry trees in the north-western US occurred at sites with different tree characteristics from those at randomly selected sites (Stewart *et al.* 1999). Damage to crops may be concentrated around the edge of a field. These empirical results suggest a fourth principle of patterns and processes of wildlife damage, **damage variation** – damage varies spatially and temporally (Table 2.1).

The variation in damage can be described by the frequency distribution of damage, i.e. the number of sites (or times) with different levels of damage (Fig. 2.5). There is a range of possibilities in the shape of the frequency distribution, for example log-normal shape has been observed in the distribution of damage by rats to rice in parts of the Philippines (Buckle *et al.* 1984) and negative exponential shape in damage by birds to grapes in Texas (Johnson *et al.* 1989).

To estimate the relationship between damage and pest density (Figs 2.1 and 2.2) the underlying variation in damage (i.e. the frequency distributions of pest damage, Fig. 2.5) must be taken into account. If we

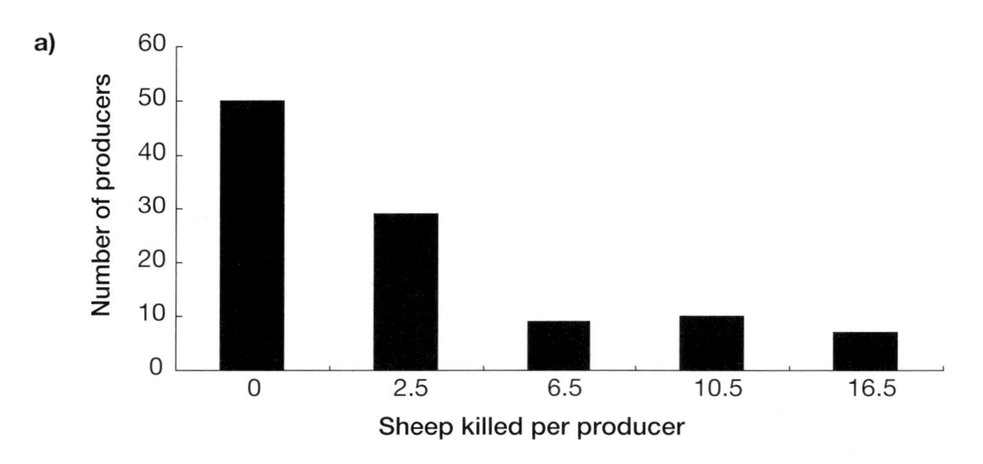

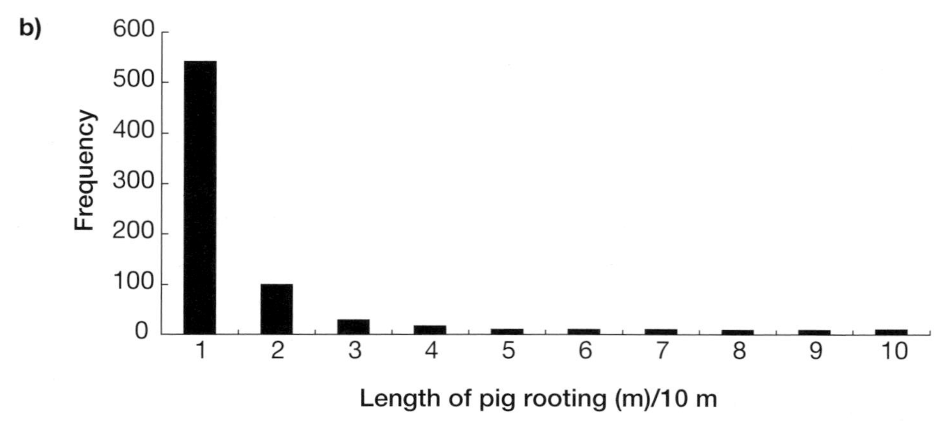

Figure 2.5: Examples of frequency distributions of damage. (a) Livestock killed, per producer, by canine predators such as coyotes, in Kansas (after Robel *et al.* 1981). (b) Ground rooting by feral pigs (after Hone 1988a).

randomly select sites for study, there are likely to be a lot of sites showing little or no damage. If those sites are used to evaluate a relationship between damage and pests, the likely result is that no relationship would be detected (and a type II error has occurred). However, if sites were chosen non-randomly to include a broad range of pest densities and damage, there is a higher likelihood that a relationship would be detected. There is no difference in the existence of a relationship, only a difference in how sites are selected and hence whether the relationship is detected.

The **damage variation** principle is illustrated in the following two worked examples, the first relating to a production issue and the second to a conservation issue. The examples also illustrate other damage principles, as noted below.

Conclusion

This chapter has described common principles in the patterns and processes of wildlife damage. There are many connections between topics in wildlife damage and those in population ecology, and much more can be learned by looking at the field in this way. The next chapter examines general patterns and processes in wildlife damage control, and discusses connections to related topics in wildlife harvesting and ecology.

Worked examples

1 Production

This worked example illustrates the **damage extent, damage response determinants** and **damage variation** principles (Table 2.1). Assume you are a scientist employed to determine whether rat damage to crops varies across fields and why it varies. A large number of quadrats are used to estimate damage. They are sampled randomly without replacement within each of 16 equal-sized fields of one type of crop, for example rice or wheat. Table 2.4 shows the mean levels of damage per field and rat abundance. Yield is estimated within fenced rat exclusion quadrats in each field. What is the mean damage, and its variance, across the fields? Was there a threshold rat abundance before damage occurred? Describe the characteristics of the study relative to the types of studies in Table 1.2 (p. 11). Is there evidence of relationships between damage and rat abundance, between yield and damage, and between yield and rat abundance?

Table 2.4: Rat damage and crop yields across a range of rat abundance. Data are hypothetical but based on a real field study

Field	Damage (kg/ha)	Rat abundance	Yield with no damage (kg/ha)	Yield with damage (kg/ha)
1	0	0	550	550
2	15	5	600	585
3	44	15	450	406
4	60	20	500	440
5	40	28	650	610
6	82	30	700	618
7	123	41	550	427
8	79	44	630	551
9	155	55	480	325
10	130	63	570	440
11	168	72	530	362
12	210	83	600	390
13	160	87	690	530
14	240	100	520	280
15	180	105	490	310
16	211	108	500	289

Solution

The mean damage across fields is 118.6 units (kg/ha) and its variance is 5594.1. The ratio of the variance to the mean is expected to be 1.0 (Krebs 1999) if damage is distributed at random. The observed ratio is 5594.1/118.6 = 47.2, which is highly significantly different from the expected value of 1.0. The statistical test used is a t-test (Manly 1992). The result supports the **damage variation** principle. The temporal component was not tested. The study is a classical experiment as shown in cell 1 of Table 1.2 (p. 11).

The damage across fields was positively related to rat abundance (Fig. 2.6a). The regression of damage (y) and rat abundance (x) was highly significant ($F = 128.54$; df = 1,14; $P < 0.001$) with the equation being:

$$\text{damage} = 13.12 + 1.97(\text{rats}) \quad (2.1)$$

The coefficient of determination (r^2) was 0.90, which is very high, as the maximum value can be 1.0. The result illustrates the **damage extent** principle. The intercept (13.12) has a standard error (SE) of 11.10; the t-test, measuring the difference of the intercept from zero, had a t value

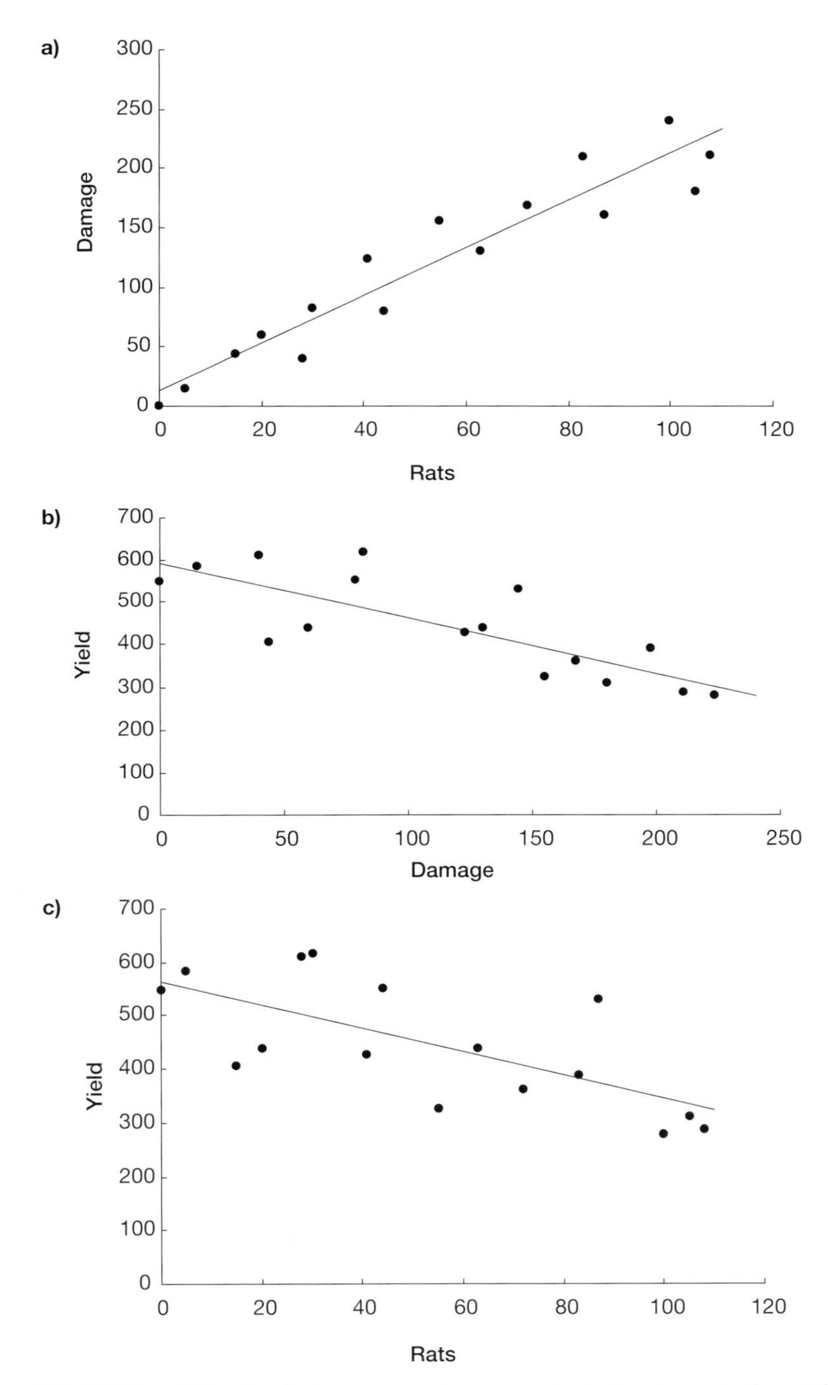

Figure 2.6: (a) The relationship between crop damage and rat abundance in the worked example (equation 2.1). (b) The relationship between crop yield and rat damage in the worked example (equation 2.2). (c) The relationship between crop yield and rat abundance in the worked example (equation 2.3).

of 1.18 (= (13.12 – 0)/11.10) and was not significant ($P > 0.05$). The results support the null hypothesis of the **damage extent** principle (no threshold pest abundance).

Crop yield decreased as damage increased (Fig. 2.6b). The observed results show a significant negative linear regression ($F = 20.19$; df = 1,14; $P < 0.001$) between yield and damage, with the slope of the regression estimated to be -1.19. The coefficient of determination (r^2) was 0.59. The equation was:

$$\text{yield} = 586.00 - 1.19(\text{damage}) \qquad (2.2)$$

The 95% confidence interval of the slope was –1.76 to –0.62. That interval includes a slope of –1.0 which would occur if no compensation occurred. Hence there is no significant evidence of compensation. The fitted regression equation predicts that yield would be 586.00 (= 586.00 – (1.19 × 0)) kg/ha when there was no damage. The fitted regression equation predicts that yield would be 0 kg/ha when damage was 492 (= 586.00/1.19). The estimated damage (492) greatly exceeds observed values (maximum 240), so we should be cautious about making the extrapolation.

Crop yield decreased significantly as rat abundance increased (Fig. 2.6c) ($F = 12.89$; df = 1,14; $P < 0.005$) with the coefficient of determination (r^2) being 0.48. The equation was:

$$\text{yield} = 563.92 - 2.23(\text{rats}) \qquad (2.3)$$

The fitted equation is a specific example of the general relationship between yield and pest abundance, illustrated in Figure 1.3b (p. 8). That relationship was explored further by Hone (2004). The fitted regression equation predicts that yield would be 563.92 (= 563.92 – (2.23 × 0)) kg/ha when there were no rats. It predicts that yield would be 0 kg/ha when rat abundance was 253. The estimated abundance of rats (253) greatly exceeds observed values (maximum 108), so again we should be cautious about making the extrapolation. This exercise illustrates patterns in many studies of rodent damage to rice crops in south-east Asia (Singleton *et al.* 1999a) and the effects of trapping on the rodents (Singleton *et al.* 1999b).

2 Conservation

This worked example illustrates the **damage response determinants** and **damage variation** principles. In a grassland conservation reserve,

Table 2.5: Plant species richness at differing levels of ground rooting by feral pigs

Quadrat	% soil rooted	Plant species richness
1	0	6
2	0	5
3	0	5
4	0	4
5	25	6
6	25	4
7	25	5
8	25	4
9	50	5
10	50	4
11	50	3
12	50	4
13	75	3
14	75	2
15	75	3
16	75	2
17	95	2
18	95	1
19	95	2
20	95	1

Source: Modified from Hone (2002)

feral (also called wild) pigs actively root up the ground to feed on plant roots and soil invertebrates, such as earthworms. You are a natural resources manager asked to review a study, part of which investigated the short-term effects (one week) of the pig rooting. The data from 0.25 m^2 quadrats are shown in Table 2.5. The scientist who provided the data had randomly sampled areas so that the rooted percentage of each plot was balanced across five levels of pig rooting (0%, 25%, 50%, 75% and 95%). It is known that when all of an area is disturbed it is converted to bare soil, and no plant species occur as living above-ground plants. The study demonstrated that as the extent of pig rooting increased, plant species richness declined significantly in a linear manner.

Linear regression showed a significant decline in species richness (S) as the percentage of a quadrat rooted ($\%R$) increased (F = 54.95; df = 1,18; $P < 0.001$) (Fig. 2.7). The coefficient of determination (r^2) was 0.75. The least squares equation of best fit was:

$$S = 5.43 - 0.04(\%R) \tag{2.4}$$

The relationship is expected to be curved (concave down), based on related topics in ecology. Why is it expected to be curved? What is an equation for the curve? What are the related topics in ecology? Given the experimental design, which type of study in Table 1.2 (p. 11) best describes the features of this study?

Solution

The linear regression equation predicts that species richness will decline to zero when the percentage of a plot with pig rooting is 136%. This is clearly an unrealistic overestimate, as $\%R$ has a maximum value of 100%.

The species area curve ($S = cA^z$) describes the relationship between species richness (S) and the area (A) of an island or habitat (MacArthur & Wilson 1967; Krebs 2001). It can be fitted to the data by replacing A with the area of unrooted (intact) vegetation (A_0), estimated as

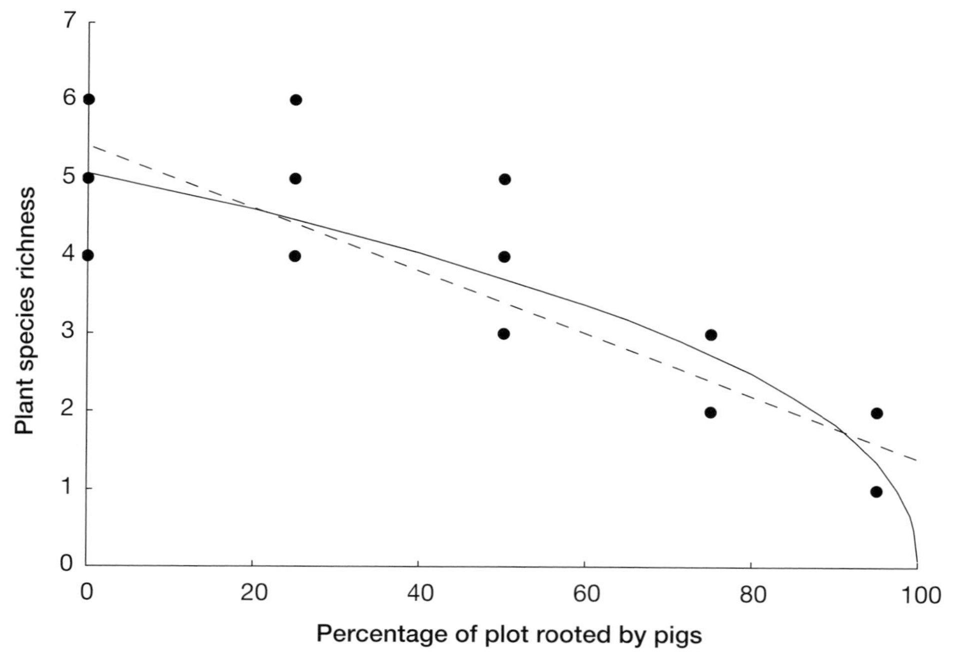

Figure 2.7: The relationships between plant species richness and percentage of each quadrat disturbed (rooted) by feral pigs. The dashed line shows a linear regression (equation 2.4) fitted to the data and the solid line shows a curved regression (equation 2.7).

$A_0(1 - (\%R/100))$. When no ground is rooted then $\%R = 0$, so $A = A_0$. When all of a quadrat is rooted then $\%R = 100$, so $A = 0$. The revised equation is:

$$S = c(A_0(1 - (\%R/100)))^z \tag{2.5}$$

The equation is fitted to the data, after transformation of the data to common logarithms (logarithms to base 10), to convert the species–area curve to a linear regression. That equation is:

$$\log S = \log c + z \log(A_0(1 - (\%R/100))) \tag{2.6}$$

Appendix 1 provides more details, if you would like assistance with some of the mathematics. The fitted equation is highly significant ($F = 74.46$; df = 1,18; $P < 0.001$), with a coefficient of determination (r^2) of 0.81 (Fig. 2.7). The equation is:

$$S = 5.06(1 - (\%R/100))^{0.44} \tag{2.7}$$

which predicts that species richness will decline to zero when the percentage of a plot with pig rooting is 100% ($\%R = 100$). This prediction agrees with the resource manager's experience. Conversely, when there is no rooting ($\%R = 0$) then $S = 5.06$. The coefficient (0.44) is a bit higher than expected in a species area relationship in island biogeography. Such coefficients are often higher when data are collected from very small plots (Krebs 2001). The curved regression is a slightly better fit, statistically speaking, than the straight line.

Aspects of the topic can be explored further and directly linked to related topics in ecology. The straight line and the curve are examples of alternative models, or multiple hypotheses (Chamberlin 1965). A requirement of models is that they make biological sense (Burnham & Anderson 1998, 2002). The straight line does not make sense ecologically and underestimates the effect on species richness when a high percentage of a quadrat is rooted over. The linear regression assumes that each time a feral pig roots up a particular size area of the ground there is an equal reduction in species richness. This is the same as saying the slope is constant. That would happen only if each plant species occurred only once on each quadrat, which is very unlikely. The expected dispersion pattern of a species is clumped (Caughley & Sinclair 1994) and the expected frequency distribution of relative abundance of species is highly positively skewed (e.g. negative exponential or log-normal) (Krebs 2001).

The curve is expected as each plant species would usually occur more than once on any quadrat. Hence, if a feral pig roots up part of a quadrat usually no species would disappear from the quadrat, as they would be represented by other individuals of the same species elsewhere in the quadrat. However, as pig rooting becomes more extensive a species may disappear. If all the quadrat is rooted over then all plants have their above-ground parts removed and may be killed.

If the feral pigs were highly selective in which plants they rooted up, a different relationship in Figure 2.7 would be expected. For example, if they selectively rooted up the rarest plant species, then the second rarest species and then the third rarest, the relationship would be a curve like that described by a negative exponential equation. Species richness would decrease rapidly with small increases in pig rooting. The data show no evidence of such selective foraging. The negative exponential equation would be another alternative model, in the sense of Burnham and Anderson (1998, 2002).

The pig rooting may be a form of disturbance but the study does not fully test the intermediate disturbance hypothesis (Grime 1973; Connell 1978) as there is no temporal component to the disturbance – only the short-term effects are studied. The long-term disturbance effects could be tested in a related study. A similar form of modelling of the effects of habitat destruction using the species–area curve was described independently by Wilson (1992) and Tilman and Lehman (1997). The study is an example of a classical experiment (Table 1.2). Further details on the worked example are given by Hone (2002).

3

Generalities in controlling wildlife damage

Introduction

In deciding to control a pest such as a rodent, a wildlife manager must make many decisions: what type of trap to use; what type of bait; where to place the trap; when to use it; how to hide it (if necessary) so that rodents don't avoid it; when to check it; what to do with captured rodents; how to avoid capturing other species and what to do if other species are captured; what permits or permissions are required; how many days to continue trapping before moving the trap, using a different trap or a different bait and so on.

Every method of wildlife damage control involves a similar array of decisions. Details of how to use such methods are well described in the literature, for example Dolbeer *et al.* (1994) and Conover (2002) discuss them in relation to wildlife of North America. Many important practical issues in controlling wildlife damage are described in books on, for example, control of feral horses (Dobbie *et al.* 1993) and rodents (Caughley *et al.* 1998). Across the broad range of species, control methods, locations and times that have been studied, have any generalisations been identified or could they be identified? As discussed in Chapter 1, using generalisations instead of approaching each problem from scratch could save managers and scientists a lot of resources, time and money.

Other fields such as ecological theory may also offer useful insights. For example, principles of harvesting can be usefully applied to wildlife damage control – harvesting and population control are analogous events (Caughley 1980; Shea *et al.* 1998). The effects of damage control on pest populations are influenced by the underlying population dynamics. A wildlife population such as a vertebrate pest population may be limited by factors such as food, predators, parasites, nest sites or competitors (Andrewartha & Birch 1954; Berryman 1999; Krebs 2001; Owen-Smith 2002; Sibly & Hone 2002; Sibly *et al.* 2003). Being aware of what controls a pest population will assist in planning control of pest damage.

Rabbit numbers may increase exponentially after rabbit control, but such an increase may not occur if the reduced population is then limited by predation. An increase in rabbit abundance would require predator control. Such an assumption has been suggested for European rabbit and its predator, the red fox, in parts of semi-arid Australia (Pech *et al.* 1992). There may also be other limiting factors such as parasites, pathogens and competitors. For example, brown-headed cowbirds (*Molothrus ater*) were controlled in order to increase abundance of a threatened bird species, the Kirtland warbler (*Dendroica kirtlandii*), in the US (summarised by Caughley & Gunn 1996). However, warbler abundance did not increase initially as some other factor apparently limited warbler abundance. By contrast, in parts of British Columbia removals of cowbirds increased the finite growth rate (λ) of song sparrow (*Melospiza melodia*) populations (Smith *et al.* 2002).

Whether a species establishes, persists and becomes a pest also depends on ecological processes. The success or failure of establishment after the introduction of ungulates and birds to New Zealand following European colonisation, for example, was positively related to the number of individuals introduced, the number of introduction events and taxa (Forsyth & Duncan 2001). Ungulates had a higher probability of successful introduction than did birds. In Australia, the success or failure of bird introductions was related to the number of individuals introduced, the number of introduction events and the area of climatically suitable habitat (Duncan *et al.* 2001). When planning to use these results for species management, it is important to remember that an introduced species may not become a pest simply by virtue of being introduced, a point related to the need to demonstrate damage (as discussed in Chapter 1).

This chapter looks at principles, or generalities, across a species (x), location (y), control method (z) and time (t) spectrum. It illustrates aspects of the framework given in Figure 1.4 (p. 9), i.e. analysing the relationships between some response variables and the level of pest control effort. Later chapters examine aspects of control specifically relating to conservation (Chapter 4), production (Chapter 5), human and animal health (Chapter 6) and recreation (Chapter 7). This chapter should be read and used in conjunction with the later chapters.

Eleven principles describing generic aspects of damage control are described and discussed. They are listed in Table 3.1 and grouped into three broad areas: pest reduction (looking at the short-term effects of control), control dynamics (focusing on the longer-term issues and variation in the level of control) and economics of control (which centres on the costs and benefits of pest control).

Pest reduction

A fundamental assumption in damage control is that reducing pest abundance causes a reduction in pest damage. Does this assumption generally hold true? A number of field studies have concluded it is valid. Examples of studies which reported a reduction in damage with a reduction in pest

Table 3.1: Principles relating to generic aspects of wildlife damage control

Pest reduction
3.1: Pest reduction. Reducing pest density reduces pest damage
Control dynamics
3.2: Control determinants. The number of pests controlled is proportional to the product of pest abundance and the control effort
3.3: Response to control. Pest control changes individual pests and pest populations
3.4: Pest removal rate. Reduction of pest density requires the rate of removal to be greater than the rate of increase of the pest population
3.5: Eradication conditions. Eradication of pests is possible only under very limited conditions – no immigration and a higher rate of removal than rate of increase
3.6: Control independence. The effects of control methods are not independent
3.7: Susceptibility to control. Heterogeneity exists between individuals in their susceptibility to control
3.8: Spatial scale of control. The scale of control is related to the spatial scale of damage and pest movements
Economics of control
3.9: Control benefits and costs. The benefits and costs of control increase as the level of control increases
3.10: Marginal response. Each unit reduction in pest damage and abundance is more expensive than the previous unit reduction
3.11: Substitution. One method of damage control can be substituted for a different method if there are higher benefits for a given cost

abundance include one on rodent damage to crops in Rajasthan, India (Advani & Mathur 1982). Significant reductions in house mouse abundance and damage were reported after mouse control in southern Australia (Mutze 1993). In Norway, control of wolverine reduced sheep predation in the year of the control but not in subsequent years (Landa *et al.* 1999). This occurred even though predation was not significantly related to wolverine abundance. Coyote control reduced predation of sheep in parts of Utah and Idaho (Wagner & Conover 1999).

These results suggest a generalisation, the **pest reduction** principle, that reducing pest density reduces pest damage (Table 3.1). While somewhat obvious, this is a fundamental principle in damage control with vertebrate pests. Two related relationships are that an increase in damage causes a reduction in yield, and that an increase in control effort causes an increase in yield (see Fig. 1.4b, p. 9). The damage yield relationship (see Fig. 2.1, p. 17) was examined in the **damage response determinants** principle (see Table 2.1, p. 16).

Not all studies produce such clear results. Some studies have demonstrated a lack of significant changes in damage after pest control. One study reported a significant change (reduction) in rabbit abundance after rabbit control, but no significant change in pasture biomass, in central Australia (Foran *et al.* 1985). Similar studies in southern Australia reported significant changes in rabbit abundance (Williams & Moore 1995) but few changes in pasture biomass and composition (Brown 1993) after rabbit control. Black rat (*Rattus rattus*) abundance in a part of Hawaii declined significantly after rat control but the yield of macadamia nuts did not increase (Tobin *et al.* 1993). Manipulation of the habitats of black rats adjacent to macadamia plantations in Queensland changed the level of rat damage and its spatial pattern across rows of macadamia trees (White *et al.* 1998). The reasons for any non-significant results need to be examined to see if type II errors have occurred (see Table 2.3, p. 23). If such errors have not occurred, the fundamental assumption behind the **pest reduction** principle may need to be re-examined.

If the relationship between pest abundance and damage (Figs 2.2 a and c, p. 21) is linear then each unit reduction in abundance will cause the same unit reduction in damage. If the relationship between pest abundance and damage is a curve, such as concave down (as in Fig. 2.1, p. 17), then the effects on damage of reducing pest abundance will vary depending on where on the curve pest abundance moves from and to.

A potential method of damage control is the use of repellents. Repellents act by making pests avoid whatever it was they previously damaged. If the pest density itself is not changed then any observed reduction in damage resulting from use of repellents corresponds to a change in the slope of the relationship between pests and damage. Estimating experimentally the effects of repellents on the density–damage relationship involves comparing the slopes of the estimated regressions, i.e. the regression slope when repellents have been used compared with the slope when repellents have not been used. The regression slope with repellent use is expected to be less than the regression slope with no repellent use. A test of slopes by analysis of covariance (Snedecor & Cochran 1967) can be used.

Control dynamics

Pest populations' responses to repeated pest control need to be recognised and incorporated into ongoing control actions. Furthermore, levels of pest control can change over time due to changes in damage, logistics, resources for control and other such variables. This section introduces principles in damage control that focus on longer-term issues and variation in the level of control.

Control determinants

Results of studies in harvesting and predation ecology suggest a key relationship that can be applied in wildlife damage control. In harvesting studies (Begon *et al.* 1996; Krebs 2001) the number of animals harvested (n) is proportional to the product of the number available to be harvested (N) and the effort made in harvesting (E). This can be expressed as an equation:

$$n = a\,N\,E \tag{3.1}$$

where a is the proportionality constant. The equation is called the surplus yield model (Begon *et al.* 1996; Krebs 2001) and in predator–prey ecology it is called the functional response (for details see Chapter 2). The relationship in this equation can be estimated by regression analysis of the relationships between n, N and E. Separate relationships (n vs N, n vs E) can be estimated by linear regression and the overall relationship (n vs N, E) by multiple regression. The assumed linear relationships can be extended to estimate curved relationships (the functional response of pest operators to a range of pest abundances) (Hone 1994a) such as by non-linear least squares regression.

A pest control operator is analogous to a harvester or predator searching for and catching prey. Many decisions are the same, such as where to search, how long to search an area (patch) and when to quit and move to another area (Charnov 1976; Hone 1990, 1994a). In wildlife damage control, a form of equation 3.1 has been used to evaluate aspects of the relationship between the number of pests removed and pest abundance for shooting (Hone 1990; Choquenot *et al.* 1999) and hunting (Caley & Ottley 1995). Figure 3.1 shows an example of a non-linear version of the relationship in equation 3.1, relating to shooting feral pigs.

With equation 3.1 as its basis, the **control determinants** principle states that the number of pests controlled is proportional to the product of pest abundance and the control effort (Table 3.1). The relationship in equation 3.1 explains the framework relationship in Figure 1.4 (p. 9), where responses in variables such as biodiversity, yield and disease cases increase in response to increased control effort because of changes in the number of pests controlled. In any given location an increase in control effort often occurs along with increased duration of the control action.

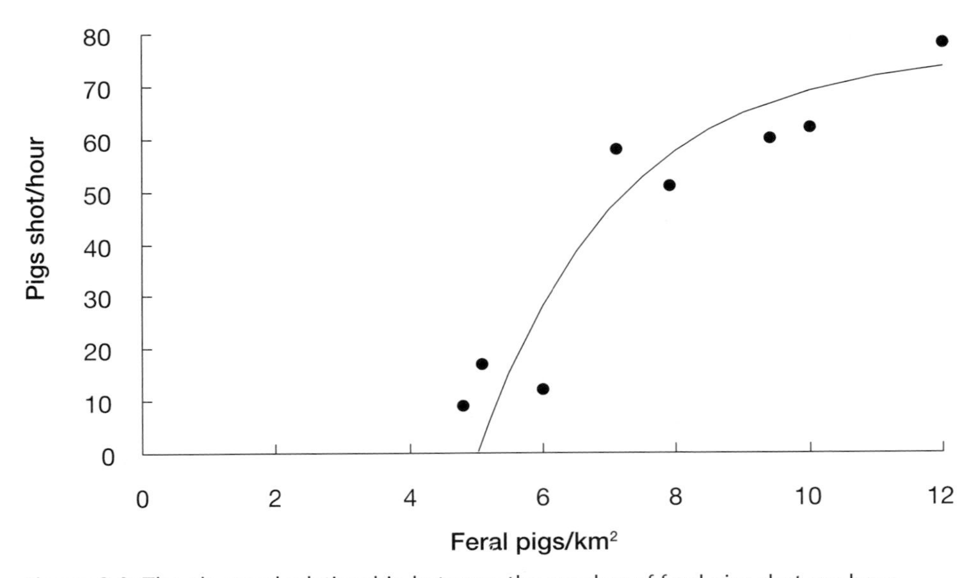

Figure 3.1: The observed relationship between the number of feral pigs shot per hour and feral pig population density (pigs/km^2) in north-western New South Wales. The solid line shows the fitted regression. The equation fitted here was $y = 76.282(1 - e^{-(x - 5.023)/2.115})$.

Source: Modified from Choquenot *et al.* (1999)

Responses to control

Ongoing pest control may cause a variety of changes in areas such as:

- behaviour of individual pests;
- genetic changes;
- demographic changes;
- changes in movement patterns and related population dynamics.

In other words, pests and pest populations respond to control efforts. The **response to control** principle states that pest control changes individual pests and pest populations (Table 3.1). It is closely linked to the **damage response determinants** principle. The latter refers to a response by what is being damaged, such as a crop, while the former refers to a response by the pest population itself. The framework relationships in Figure 1.4 (p. 9) do not show the responses by pests; however, the patterns in biodiversity, yield and disease cases would be partly generated by such responses.

The behaviour of individual pests may change in response to pest control. If that occurs, the effectiveness of the control methods can change over time. For example, shooting feral pigs from a helicopter can be affected by a few pigs learning to hide in patches of dense vegetation (Saunders & Bryant 1988). The phenomenon involves individuals of a prey species modifying their behaviour to avoid predation. Other examples are the development of bait shyness caused by previous illness, described for brushtail possums in New Zealand feeding on sublethal cyanide bait (Warburton & Drew 1994; Morgan *et al.* 2001) or bait containing 1080 poison (Ross *et al.* 2000), and bait avoidance in rats in oil palm crops in Malaysia (Wood & Fee 2003).

A form of behaviour change may correspond to a changed state of individuals in relation to pest control. For example, before poisoning occurs individuals can be susceptible, once poisoning occurs but before signs develop they can be called 'latents', after they develop signs of poisoning they can be called 'exposed' and behaviour changes may occur (Hone 1992). The same sequence of changes occurs with individual hosts and microparasites and macroparasites (Anderson & May 1979, 1991). An analysis of feral pig trapping in southern Australia (Choquenot *et al.* 1993) was based on a change of state hypothesis.

Genetic changes may occur with natural selection in response to pest control. Classic examples of such changes are the evolution of rodents' resistance to anticoagulant pesticides (Greaves 1985; Gill *et al.* 1994), evidence of resistance by European rabbits to compound 1080 poison (Twigg *et al.* 2002) and the evolution of the virulence of myxoma virus to rabbits in Australia, Britain and France (Fenner & Fantini 1999). Other pest control methods such as trapping, shooting or other pesticides can also cause changes in the genotype of pest populations.

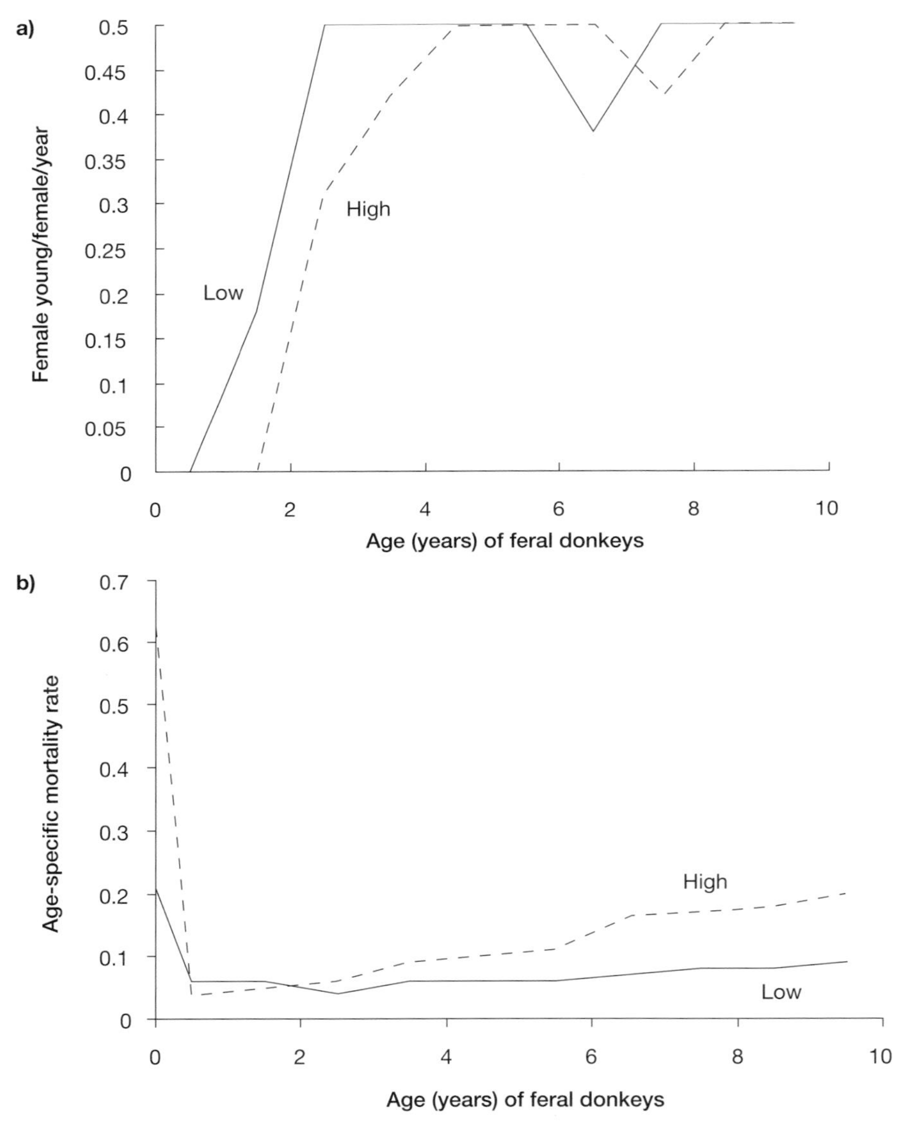

Figure 3.2: (a) Observed differences in age-specific fecundity (m_x) of feral donkeys in a low-density and a high-density population (redrawn from Choquenot 1991). (b) Observed differences in age-specific mortality rate (q_x) of feral donkeys in a low-density and a high-density population (redrawn from Choquenot 1988).

There may also be demographic changes in a population in response to control. Reduction of pest abundance may cause a compensatory change in pest mortality and/or fecundity, or other population parameters. When abundance is reduced then competition for resources is reduced, reflecting Berryman's (1999) third principle of population dynamics – that individual organisms may compete for resources when their population becomes large. Feare (1991) considered that compensatory changes after pest control were one of the two main reasons for failed population reduction strategies to control bird pest problems. The second reason was incomplete exposure of the pest population to control, discussed below in the **spatial scale of control** principle.

A number of specific examples of demographic changes in response to control have been described in the literature. During efforts to eradicate feral goats (*Capra hircus*) from Raoul Island in the south Pacific Ocean, the number of foetuses increased from 0.96 to 1.70 kids per female per year (Parkes 1990a). The suggested reason was an increase in per capita food supply as goat abundance decreased. Donkeys (*Equus asinus*) in a low-density post-control population had higher fecundity (Fig. 3.2a) and lower mortality (Fig. 3.2b) than donkeys in a high-density population that had not been subject to lethal control (Choquenot 1988, 1991). Possible compensatory changes in population parameters after pest control include decreases in mortality and increases in fecundity as pest density decreases (Fig. 3.3).

Compensatory changes in survival and/or recruitment could occur in response to control such as sterilisation (Barlow *et al.* 1997). Sterilisation of some European rabbits in southern Australia increased juvenile survival and survival of infertile adult females (Twigg & Williams 1999; Twigg et al. 2000). Higher survival rates occurred in ricefield rats (*Rattus argentiventer*) in Indonesia after sterilisation (Jacob *et al.* 2004). The population growth rate (r) of white-tailed deer (*Odocoileus virginianus*) in Maryland decreased with higher levels of sterilisation (Rutberg *et al.* 2004). The results suggest that about 60% of deer need to be sterilised to stabilise population growth ($r = 0$). Sterilisation of brushtail possums in New Zealand reduced per capita recruitment rates; however, immigration compensated for such reductions leading to stable populations ($\lambda = 1$, $r = 0$) (Ramsey 2005).

In New Zealand populations from which some ferrets (*Mustela furo*) were removed, juvenile survival increased compared with populations from which no ferrets were removed (Byrom 2002). Juvenile survival was negatively density-dependent (highest survival was at lowest density, and lowest survival at highest density). In general the effects of density on births and survival can be redrawn (Fig. 3.4) to show the combined effects of lethal control and fertility control (Hone 1999).

Movement patterns of pests may change in response to population control. When population control reduces pest abundance locally, animals in nearby areas may change their movements in response to the reduced abundance and increased per capita resources. This is called the 'vacuum effect' (Efford *et al.* 2000). A study of brushtail possums in New Zealand demonstrated a vacuum effect on local movements of possums in areas adjacent to an area of previous possum control (Efford *et al.* 2000). The effects on movement patterns in response to control is a general issue relevant to many pest species and needs to be incorporated into the early stages of planning any wildlife damage control.

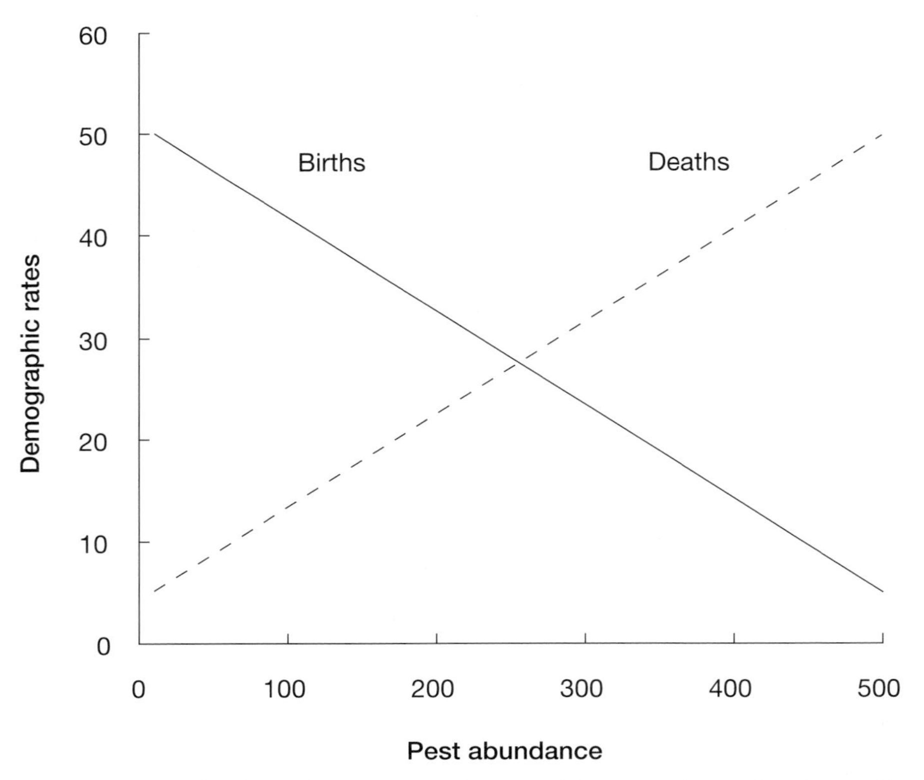

Figure 3.3: Possible relationships between demographic rates and pest abundance. The birth rate (solid line) may increase as pest abundance is reduced and the mortality rate (dashed line) may decrease as abundance decreases in response to pest control.

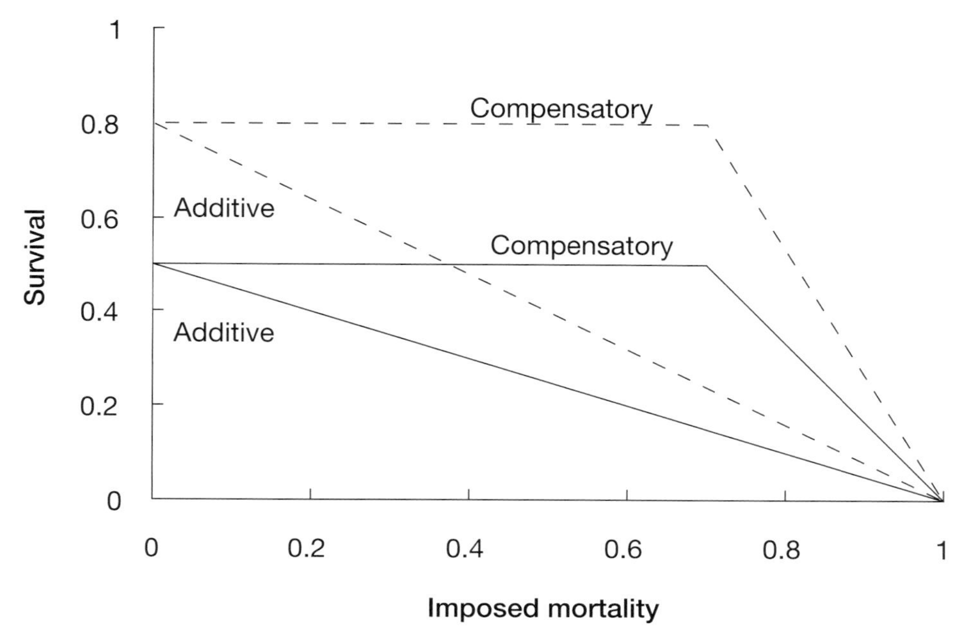

Figure 3.4: A model of additive (simple linear) and compensatory (compound linear) changes in survival of a vertebrate pest population in response to imposed mortality (solid lines) and fertility control (dashed lines).

Source: Modified from Hone (1999)

In population ecology the issue of movements is incorporated into population dynamics in discussions of source populations and sink populations. A source population is a net exporter of individuals and a sink population is a net importer (Pulliam 1988; Thomas & Kunin 1999; Kadmon & Tielborger 1999). Sustained control can convert a source population to a sink population by increasing the mortality rate so that it is higher than the birth rate. The population could then only be sustained by continued immigration, and would be a sink population. Hence, control of a source population may have a greater effect on pest abundance than control of a sink population. A source population can increase in the absence of pest control but a sink population cannot, so application of pest control to a source population should reduce pest abundance overall. If pest control continues at a high rate then the source population may be converted to a sink population.

When movement is prevented into or out of very small populations there may be changes in population dynamics. If an animal is rare and many animals in that population cannot find mates (the Allee effect) the rate of increase (r) of the population can decline (Fig. 3.5). When abundance goes below the lower equilibrium, k, abundance declines rapidly to extinction (Grossman & Turner 1974; Stephens & Sutherland 1999; Courchamp *et al.* 1999a; Berryman 1999; Dennis 2002). The condition for such decline may be created by isolation of a very small population, as has been suggested in wildlife damage control for enclosure fencing of feral pigs in Hawaii Volcanoes National Park (Hone & Stone 1989). In pest control it is often assumed that small isolated pest populations will not go extinct naturally and must be totally eradicated; however, the Allee effect hypothesis suggests such natural extinction can occur. The relationship demonstrated in Figure 3.5 may also be generated by a different cause, such as predation (Sinclair *et al.* 1998; Berryman 1999), so it is necessary to distinguish between alternative explanations for the relationship.

In addition to changes in a pest population in response to control, control efforts may create unintended consequences for non-target

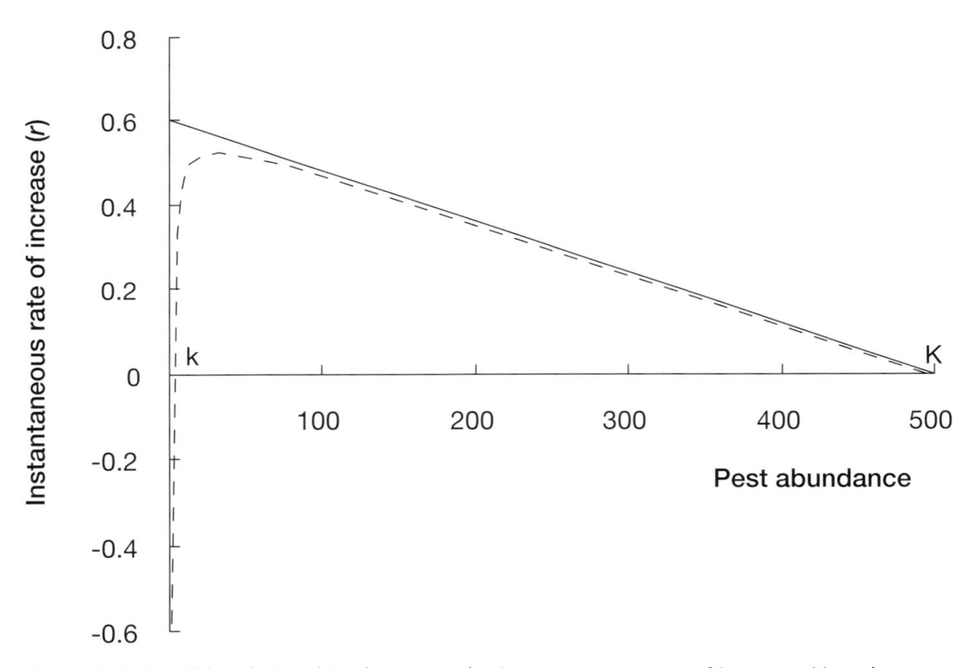

Figure 3.5: Possible relationships between the instantaneous rate of increase (r) and population density. The linear relationship (solid line) describes logistic growth and the curved (dashed line) relationship could occur when, for example, pests have difficulty finding mates at very low population density (the Allee effect). There is an upper equilibrium abundance, K (= 500), and a lower equilibrium abundance, k (= 2), and the annual intrinsic rate of increase (r_m) equals 0.6.

Source: Modified from Courchamp *et al.* (1999a)

species. Control of one pest species can produce a change in the pest status of other species, a phenomenon called 'pest replacement'. In this book such unintended consequences are considered as part of the response to control. Many instances of pest replacement have been reported. Control of feral water buffalo (*Bubalus bubalis*) in part of the Northern Territory was followed by a doubling in abundance of feral pigs (Corbett 1995). In southern Australia rabbit abundance increased after control of red fox (Banks *et al.* 1998). The control of rabbits may have implications for bird conservation because of changes in cat predation, as examined by mathematical modelling (Courchamp *et al.* 1999b). In oil palm crops in Malaysia, *Rattus tiomanicus* was replaced by *R. rattus diardii* over six years (Wood & Fee 2003).

Walker and Norton (1982) described pest replacement as a principle in applied ecology. The response of the second (or other) species depends on several factors:

- their niche breadth (the range of environmental conditions within which the species can survive and reproduce);
- the effects of interspecific competition (competitive exclusion is removed);
- the effects of apparent competition (the effects of shared parasites may limit one species more than the other species).

A second aspect of unintended consequences relates to the effect of pest control on non-target species. When an individual of a species other than the target pest species is injured or killed, it is called a 'non-target effect'. The classic example is death of a non-target species after baiting. The effect of rodenticides, targeted at rats and mice, on barn owls (*Tyto alba*) in Britain (Newton 1998) was a non-target effect. Another example was deaths of stoats (*Mustela erminea*) in parts of New Zealand after baiting for possums and rats (Murphy *et al.* 1999). There are other non-target effects, for example a second species may eat poisoned bait and receive a sublethal dose that induces bait shyness, thus limiting future control options for that second species (Hickling *et al.* 1999). Similarly, trapping for a pest may accidentally trap a threatened species (Morriss *et al.* 2000).

Investigations should distinguish between non-target effects on individuals and non-target effects on populations. For example, poisoning of Norway rats (*Rattus norvegicus*) and Pacific rats (*R. exulans*) to eradicate them from an island near New Zealand had the unintended effect of killing some native birds but may not have had an effect on breeding and recruitment by the bird populations as a whole (Empson & Miskelly 1999). The likelihood of non-target species being poisoned by 1080 poison in Australia was assessed in a combination of laboratory (McIlroy 1986) and field (McIlroy & Gifford 1991) studies. Mathematical modelling has also been used (Courchamp *et al.* 1999b, c).

The 'precautionary principle' of resource management has been proposed in other fields and has been discussed by Burgman and Lindenmayer (1998) relative to biodiversity conservation. It can be restated relative to possible side effects of pest control: if the unintended consequences of damage control are large and likely to occur, then do not do damage control.

Pest removal rate

If animals are removed from the population faster than the population can naturally increase, for example faster than the intrinsic rate of increase, abundance must fall. Caughley and Sinclair (1994) described this as one of several principles of wildlife control, and Krebs (2001) as one of several principles of harvesting. The finite rate of increase (λ) of a population is simply the ratio of abundance at successive times, such as year 2 to year 1 ($\lambda = N_2/N_1$) (Caughley 1980). The instantaneous rate of increase (r) is the natural logarithm of the finite rate of increase: $r = \ln \lambda$. The maximum value of r is also called the intrinsic rate of increase, r_m. If the rate of removal continues to be higher than the rate of increase then the population goes extinct (Fig. 3.6). This is analogous to removing money from a bank account faster than the account is generating interest. The analogy is examined further in Chapter 5.

For wildlife damage control, this suggests the **pest removal rate** principle. This states that a reduction of pest abundance requires the rate of removal to be greater than the rate of increase of the pest population (Table 3.1).

The proportion (p) of animals to cull in a short time to stop population growth in that year is estimated (after Caughley 1980 and Hone 1999) as:

$$p = 1 - (1/\lambda) \tag{3.2}$$

or:

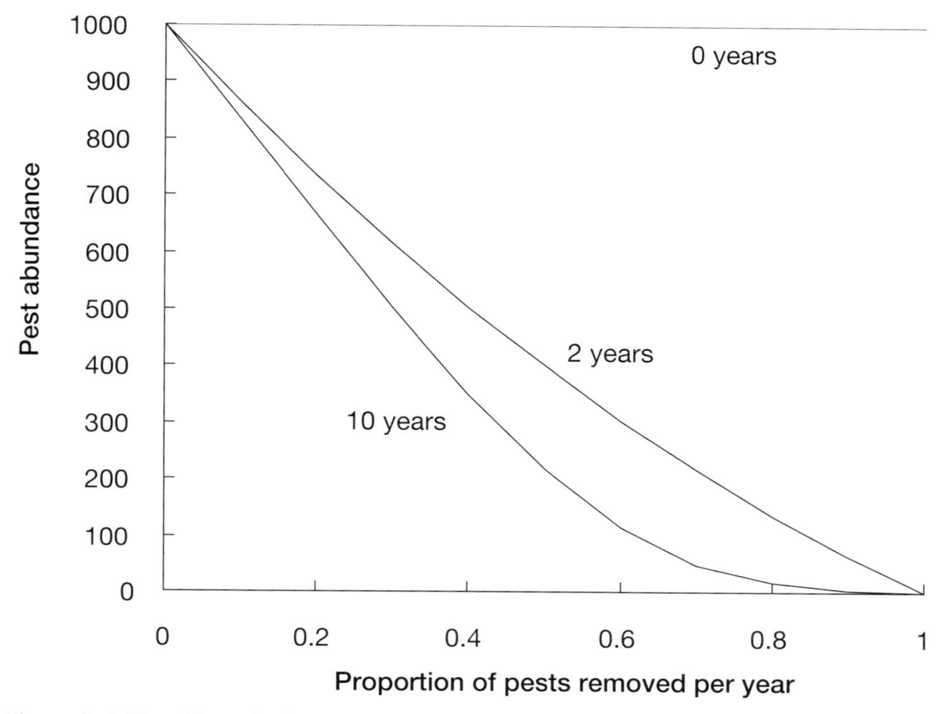

Figure 3.6: The effect of culling different proportions of pests per year from a population showing logistic growth in the absence of such culling. At the start N = 1000, the carrying capacity (K) and the intrinsic rate of increase (r_m) = 0.6 per year. The lines show the initial population, abundance after 2 years and after 10 years. Note that when 0.30 of pests are removed annually the abundance stabilises at 500 after about 20 years. If the proportion removed annually is greater than 0.3 the population goes extinct.

Source: Modified from Caughley (1980, Fig. 11.6)

$$p = 1 - (1/e^{r}) = 1 - e^{-r} \tag{3.3}$$

for years when the population increased ($\lambda > 1$ and $r > 0$). Equation 3.2 is illustrated in Figure 3.7, showing that p increases as λ increases though at a decreasing rate. There is a threshold value of λ of 1.0 below which control is not needed as the population is decreasing. The estimated values of p for several species are shown.

The classical theory of biological pest control predicts that pest abundance will decrease to a lower stable equilibrium after introduction of a biological control agent (Krebs 2001). The equilibrium hypothesis is evaluated by applying a control and seeing if pest abundance stays low at a constant level (equilibrium point) or fluctuates but shows no trend up or down. In this test, pest control must be continually applied. Trend can be assessed as the slope of the linear regression of the natural logarithms

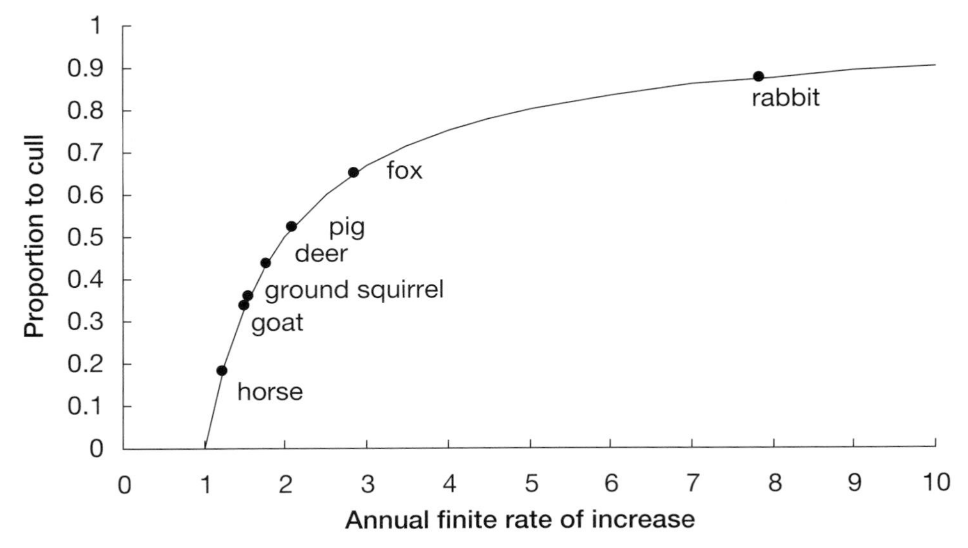

Figure 3.7: The relationship between the proportion (p) of a pest population to remove annually to stop population growth and the maximum annual finite rate of increase (λ), based on equation 3.2. The estimated value of λ for each species was: feral horse ($\lambda = 1.22\ yr^{-1}$), feral goat ($\lambda = 1.51\ yr^{-1}$), Belding's ground squirrels ($\lambda = 1.56\ yr^{-1}$), white-tailed deer ($\lambda = 1.78\ yr^{-1}$), feral pig ($\lambda = 2.10\ yr^{-1}$), red fox ($\lambda = 2.86\ yr^{-1}$) and European rabbit ($\lambda = 7.85\ yr^{-1}$).

Sources: Horse data are from Eberhardt (1987, 2002), goat data are from Maas (1997, 1998), ground squirrel data after Sherman and Morton (1984), deer data are from Nielsen *et al.* (1997), pig data are from Hone (2002), fox and rabbit data are from Hone (1999).

(logarithms to base e) of abundances over time (Caughley & Sinclair 1994). An evaluation of biological control by Murdoch *et al.* (1985) developed a non-equilibrium model, and concluded that abundance can be lowered but not to a stable equilibrium point. The quest to more closely link theory and practice of biological control continues (Murdoch & Briggs 1996; Fenner & Fantini 1999).

If pest populations fluctuate widely in a non-equilibrium manner then pest control could have an accentuated effect on pest abundance. Less population control is required to reduce pest abundance in a variable than in a constant environment. In a variable environment pest population abundance should fluctuate more than in a constant environment (Berryman 1999). We may never encounter a constant field environment, only a constant laboratory environment. In a randomly fluctuating environment the variability of harvest increases as harvesting effort increases (Beddington & May 1977). Harvesting is analogous to offtake by pest control. Sustainable harvests of magpie goose (*Anseranas semipalmata*) populations in northern Australia were predicted to be lower in a variable than in a constant environment (Bayliss 1989).

The removal rate may have a greater effect on pest abundance if control is selective rather than random. Applying control to the portion of the pest population that contributes most to the next generation may have a greater effect on abundance than random control. Individuals differ greatly in the number of young they contribute to the next generation (Clutton-Brock 1988; Newton 1989). The number of young is called lifetime reproductive success (LRS). For example, in seven species of birds 50% of young in the next generation were produced by only 3–9% of females (Newton 1995). If the individuals contributing most young could be targeted for lethal or fertility control, there may be a greater effect on pest abundance than if control occurred at random with respect to LRS.

A related aspect is that the age classes of a pest population with the highest reproductive output are those that occur shortly after sexual maturity (Fig. 3.8). Individuals in these age classes have lower fecundity (m_x) than older females but there are many more of them (higher l_x). Reproductive output is estimated as the product of age-specific fecundity (m_x) and age-specific survival from birth (l_x) where x is age. Across age classes the sum of the products ($\Sigma\, l_x m_x = R$) is called the net reproductive rate (R) (Krebs 2001). The lowest reproductive outputs are obviously in age classes before sexual maturity and in the oldest age classes (Fig. 3.8). An analysis of the sensitivity of reproductive output (the $l_x m_x$ curve) to changes in demographic parameters, such as age at first reproduction, adult survival and juvenile survival, was reported by Eberhardt (1985) in relation to a long-lived species, showing that changes in adult survival had the greatest effect on rate of increase. The sensitivity analysis was extended to many long-lived vertebrates and showed the large sensitivity of the finite population growth rate (λ) to annual adult survival (Eberhardt 2002). In contrast, in red fox populations in Britain, USA and Australia animals aged from birth to one-year-old and from one year to two-years-old had the highest sensitivity and contributed most to the finite population growth rate (λ) (McLeod & Saunders 2001).

Individual pests could be controlled before, on average, they contribute young to the next generation of pests. Such a control option involves the average age at which individual pests are controlled (lethally or non-lethally) being less than the average age at reproduction (generation

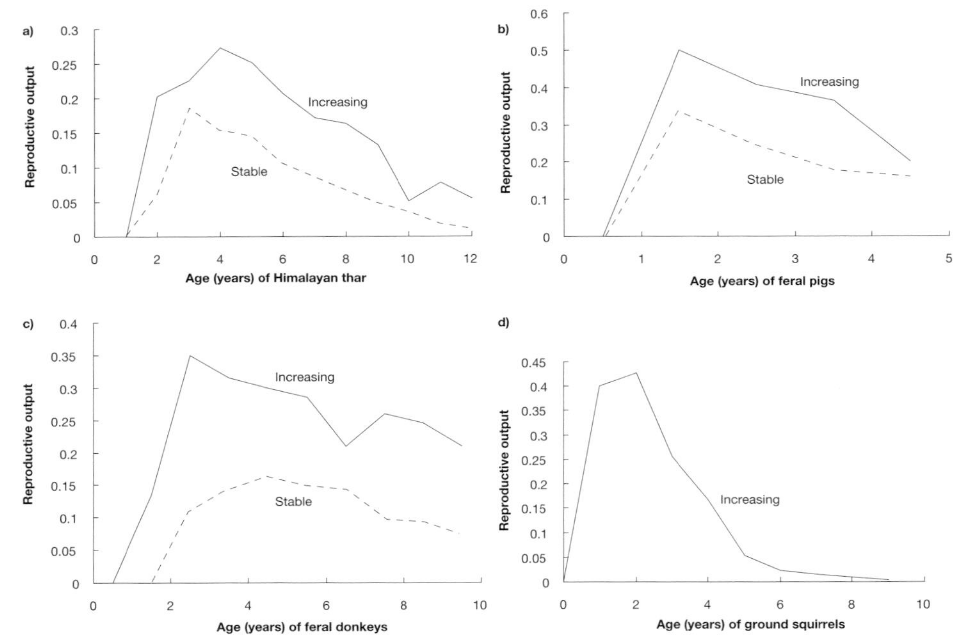

Figure 3.8: Estimated reproductive output for each age class of four mammal species: (a) Himalayan thar (*Hemitragus jemlahicus*), (b) feral pig, (c) feral donkey and (d) Belding's ground squirrel. Reproductive output is the product of age-specific fecundity (m_x) and survival from birth (l_x) for each age class. Net reproductive rate ($R = \Sigma l_x m_x$) for each population is the area under each curve. The data for Himalayan thar are for an increasing population (solid line) and a stable population (dashed line), after Caughley (1970). Data for feral pigs are for an increasing population (solid line) and a stable population (dashed line), after Giles (1980). Data for feral donkeys are an increasing population (solid line) and a stable population (dashed line), after Choquenot (1991). Data for Belding's ground squirrels are for an increasing population (Sherman and Morton 1984).

interval). The generation interval is the average age of females when giving birth to their young (Caughley 1980). An analysis, by modelling, concluded that sterilising females after they had bred was likely to have little effect on pest density (Hobbs *et al.* 2000).

The application of the **pest removal rate** principle may appear to have a methodological problem: how to know if you are removing animals faster than the rate of increase if the population has not been previously disturbed. One approach is to assume the rate of increase is equal to the intrinsic rate of increase (r_m), the maximum rate in the best of circumstances. This is a conservative approach suitable for managers who are risk-averse; the actual rate of increase (r) will usually be less than r_m. The intrinsic rate of increase (r_m) can be estimated empirically (which is better) or, if that is not possible, from general equations linking r_m and

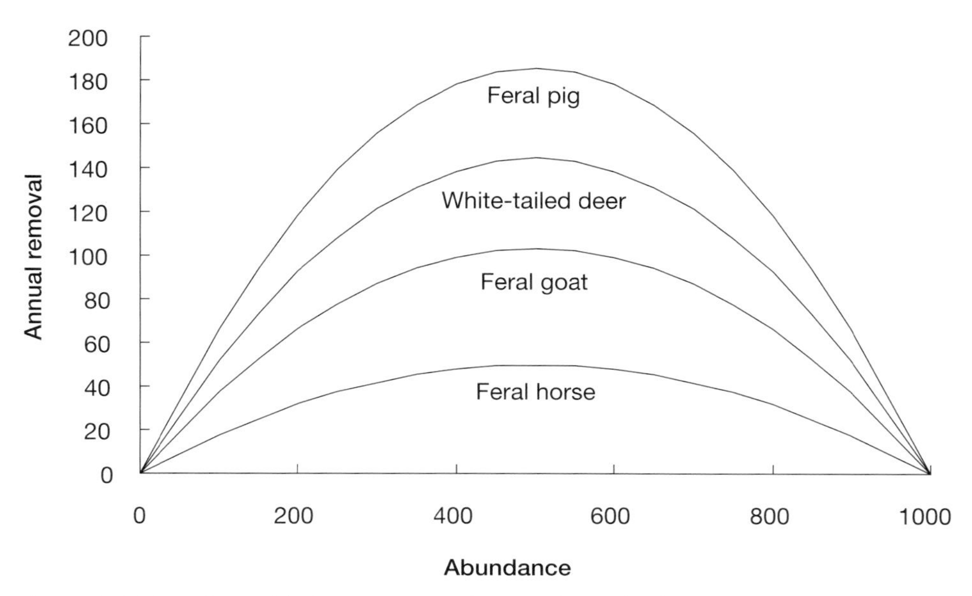

Figure 3.9: The relationship between annual removals required to maintain pest abundance at a certain level and that abundance, assuming logistic population growth and specific values for the intrinsic rate of increase (r_m). The lines are for feral horse ($r_m = 0.2\ yr^{-1}$ after Eberhardt 1987, 2002), feral goat ($r_m = 0.414\ yr^{-1}$ after Maas 1997, 1998), white-tailed deer ($r_m = 0.578\ yr^{-1}$ calculated from Nielsen *et al.* 1997) and feral pig ($r_m = 0.742\ yr^{-1}$ after Hone 2002). Carrying capacity (K) equals 1000.

bodyweight (Blueweiss *et al.* 1978; Caughley & Krebs 1983; Charnov 1993; Sinclair 1996). The number of animals to remove annually to maintain animal abundance at some level, assuming logistic growth, is related to abundance and the intrinsic rate of increase (Fig. 3.9). As r_m increases, more animals need to be removed annually.

Eradication

When the removal rate continues to be higher than the rate of increase, eradication occurs. Eradication of a pest is rarely achieved except in small areas or on islands. One successful example was the eradication of coypu (*Myocastor coypus*) in south-eastern England (Gosling & Baker 1989). Some suggested eradications may simply be interesting but unrealistic hypotheticals: only time will tell. For example, a description of a possible scenario of Australian agriculture in 2025 included the assumption that rabbits, cats and foxes had been eradicated (Derrick & Dann 1997).

Criteria for assessing efforts to eradicate pests have been developed (Parkes 1990b; Bomford *et al.* 1995; Bomford & O'Brien 1995) and partly evaluated (Rainbolt & Coblentz 1997; Hone 1999; Forsyth *et al.* 2003). Three criteria were described as essential for eradication:

1 all reproductive pests are exposed to the control efforts;
2 there is no immigration;
3 the rate of pest removal is greater than the rate of increase.

If the three conditions are met then eradication appears inevitable.

Such criteria may be unnecessarily stringent. A population may have a spatial source/sink structure (Pulliam 1988) which means that control of only the source part of the population is necessary. If the source part of the population is removed then the sink part will disappear naturally, as no individuals will move from the source to the sink and hence maintain the sink. This assumes, however, that no individuals move from the sink population to establish a new source population. As an isolated sink population will disappear naturally, not all reproductive individuals in the total population need to be exposed.

Modelling the removal of a constant percentage or proportion of a population pushes the population to extinction if the percentage is higher than a critical threshold (Caughley 1980; Caughley & Sinclair 1994). The modelling assumes control occurs at random in the pest population. Such an assumption may be unrealistic because control intensity will vary spatially and temporally (the **spatial scale of control** principle, see below) and individuals vary in their susceptibility to control (the **susceptibility to control** principle, see below).

From these empirical and theoretical results we can suggest the **eradication conditions** principle. This states that eradication of pests is possible only under very limited conditions – no immigration and a higher rate of removal than rate of increase (Table 3.1). Note that eradication may require high levels of control effort, so may require effort at the extreme right of the x axis in the framework relationships of Figure 1.4 (p. 9).

Control independence

Wildlife damage control may use many methods in the one location, such as trapping, shooting and so on. The effects of such control methods are not independent. Many research studies focus on the effects of one method, sometimes on the effects of several methods in combination. Examples of the latter are evaluations of methods of rabbit control in southern Australia (Cooke 1981; Williams & Moore 1995). In a broader pest control context, the simultaneous use of a lethal method and biological control using a lethal pathogen may result in decreased virulence of the pathogen (Carpenter 1981). Hence the effects of the lethal and the biological control methods were not independent; one had effects on the other. This modelling result was developed (Hone 1994a) as a hypothesis to suggest that the observed evolution of virulence of myxomatosis to rabbits in Australia may have been influenced by concurrent lethal rabbit control.

From such studies comes the **control independence** principle. This states that the effects of control methods are not independent (Table 3.1). For example, an 80% kill of rabbits by each of two control methods cannot give a 160% kill of rabbits. Analysis of the data from such experimental studies was suggested (Cooke 1981) to occur after logarithmic transformation, to get additivity of effects, on a logarithmic scale. The concepts of additive and multiplicative effects are described in Appendix 1. The framework relationships in Figure 1.4 are relevant to the **control independence** principle as they show the outcome of the effects of two or more control methods. However, Figure 1.4 does not show the component parts of the effects.

Susceptibility to control

Methods of pest control may differ in their effects because, for example, method 1 removes more males than females and method 2 removes older rather than younger individuals. For example, feral pigs caught by dogs in northern Australia had a sex ratio biased towards males (Caley & Ottley 1995). Brushtail possums in New Zealand (Hone 1994a) and feral pigs in Australia (Hone 1983, 2002) showed differences within populations in whether they ate poisoned bait and, if so, whether they died. An analogous concept occurs in mark–recapture studies where some estimators of abundance assume that heterogeneity does exist (M_h, M_t, M_b, M_{ht}, M_{bt}, M_{bh}, M_{htb}) or does not (M_o) in capture probabilities (Otis *et al.* 1978; Krebs 1999).

These empirical and theoretical results suggest the **susceptibility to control** principle. This states that heterogeneity exists between individuals in their susceptibility to control (Table 3.1). The framework relationships in Figure 1.4 are generated by processes that include the heterogeneity between individuals.

Spatial scale of control

Pest control is commonly applied in a non-uniform manner across an area or across time. The lack of pest control in some areas means that some or much of the pest population is inaccessible to control operations. Immigration from these untreated areas to adjacent treated areas may quickly counteract effects of pest control – the vacuum effect (Efford *et al.* 2000). The inaccessibility of some of a bird pest population to control methods was considered by Feare (1991) to be a major reason for failure of the population reduction strategy. To overcome the problems of immigration and inaccessibility, control should occur more evenly over larger areas. If attempts are made to control pest wildlife around threatened wildlife or a threatened system such as a crop or livestock, the control area needs to be of a size equal to or greater than the average home range size of the pests. The area of predator (ferret and feral cat, *Felis catus*) control in one study in New Zealand estimated the width of a buffer zone as equal to the upper 95% confidence interval of observed home range lengths (Norbury *et al.* 1998). The median dispersal distance of juvenile ferrets in another New Zealand study was 5 km, but the maximum distance was 45 km (Byrom 2002). In a study of lamb predation by red fox in southern Australia, fox control treatments (poisoning) were applied over sites larger than twice the average home range size of foxes (Greentree *et al.* 2000).

These empirical results suggest the **spatial scale of control** principle. This states that the scale of control is related to the spatial scale of damage and pest movements (Table 3.1). Scale of control here refers to the spatial extent, such as in square kilometres. If the spatial scale of control is too small it is likely that a study will not show the framework relationships in Figure 1.4.

Economics of control

The effects of wildlife damage control can be examined in monetary terms, such as the value of the costs and benefits of control. This section examines the application of several economic principles to wildlife damage control.

Control benefits and costs

As the level of pest control (the amount of bait, number of traps, length of fencing etc.) increases there are increasing benefits, such as higher yield, though they occur at a diminishing rate. Each time you add an

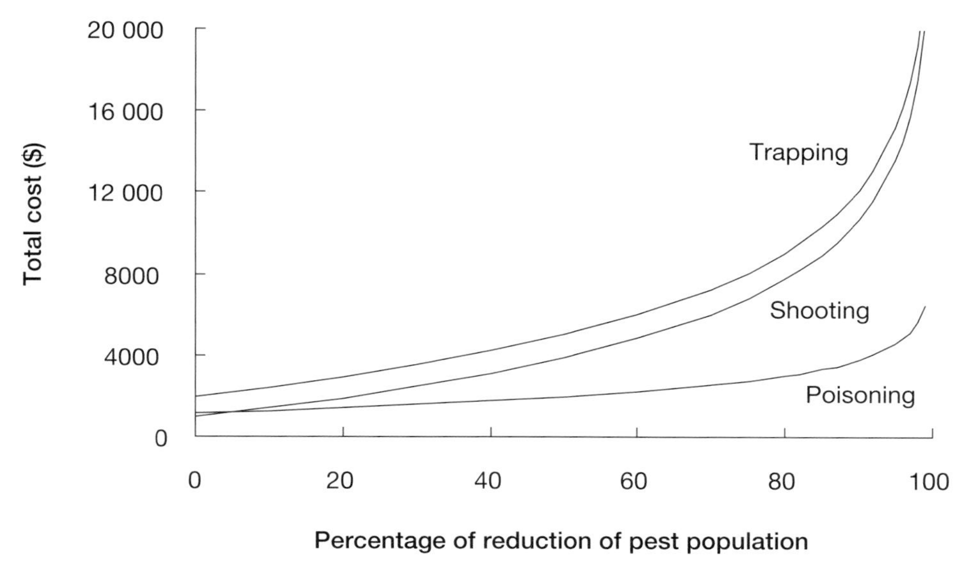

Figure 3.10: Estimated total costs of reducing the abundance of feral pigs by poisoning, trapping and shooting from a helicopter in parts of south-eastern Australia.

Source: Modified from Saunders (1988)

extra level of pest control you get an added benefit, though it is generally smaller than previous increases in pest control inputs (see Figure 1.4b, p. 9). This is the well-established economics principle of diminishing returns (Malcolm *et al.* 1996). When control efforts increase it follows that control costs increase (Figure 1.4d). The topic is examined further in Chapter 5.

As a related effect, the greater the required reduction in pest abundance, the higher the cost of control. An example is shown in the estimated total costs of achieving a specified level of reduction of feral pigs (Saunders 1988). Across each different method of control used, total costs increased at an accelerating (finite) rate as the reduction increased (Fig. 3.10). The costs of reducing abundance of European rabbits also increased at an accelerating rate (Hone 1994a).

These empirical and theoretical results suggest the **control benefits and costs** principle. The principle states that the benefits and costs of control increase as the level of control increases (Table 3.1). The principle can be examined by overlaying the framework benefit relationships (Fig. 1.4a–c, p. 9) and the cost relationship (Fig. 1.4d).

If the likelihood of damage is uncertain but it will cause major impacts if it occurs, pre-emptive damage control may be appropriate (Davis *et al.* 2004b). It is relevant to the economic assessment of the benefits and costs of control. Payoff matrix analysis in decision theory (Hone 1994a) can be used to evaluate the consequences of actions under uncertainty. This requires estimates of damage with and without pest control and estimates of the probability of such damage.

Marginal response

During a period of pest control, the number of pests captured or removed generally increases over time (as mentioned in the **control determinants** principle). By contrast, the rate of capture of new pests (marginal rate) declines. The decline in capture rate with increasing effort is a practical example of the marginal value theorem (Charnov 1976) in predator–prey theory. This effect has been demonstrated for shooting of feral buffalo (Skeat 1990) and feral pigs (Hone 1990), trapping of feral pigs (Choquenot *et al.* 1993) in Australia, and trapping of brown tree snakes (*Boiga irregularis*) in Guam (Engeman *et al.* 2000) (Fig. 3.11).

The rate of capture may decline because pest abundance has decreased but also because pests may become harder to capture (the **response to control** principle, Table 3.1). As a consequence, the cost of reducing pest abundance to a particular level increases as that abundance decreases, as shown in the framework relationship in Figure 1.3d (p. 8). Other examples have been reported for feral buffalo (Fig. 3.12a), feral goats

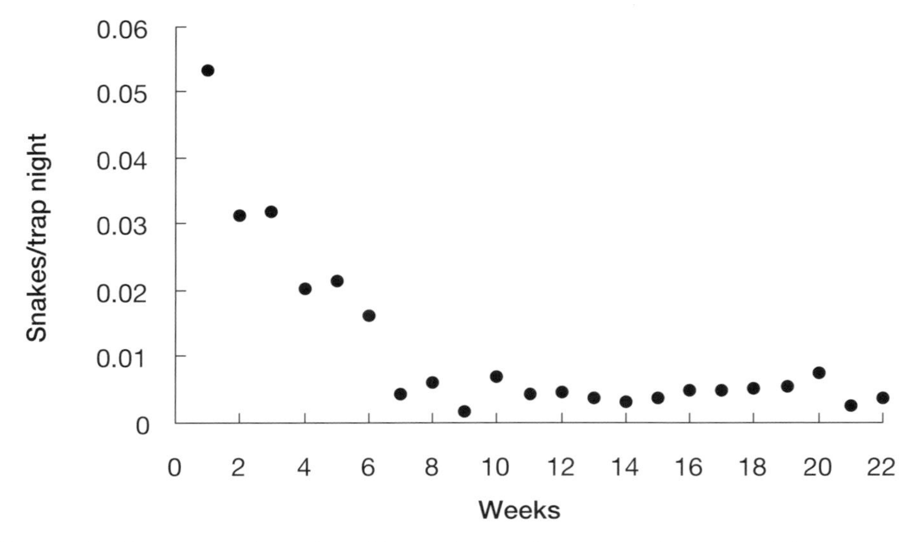

Figure 3.11: The observed number of brown tree snakes captured per trap night over weeks of trapping.

Source: After Engeman *et al.* (2000)

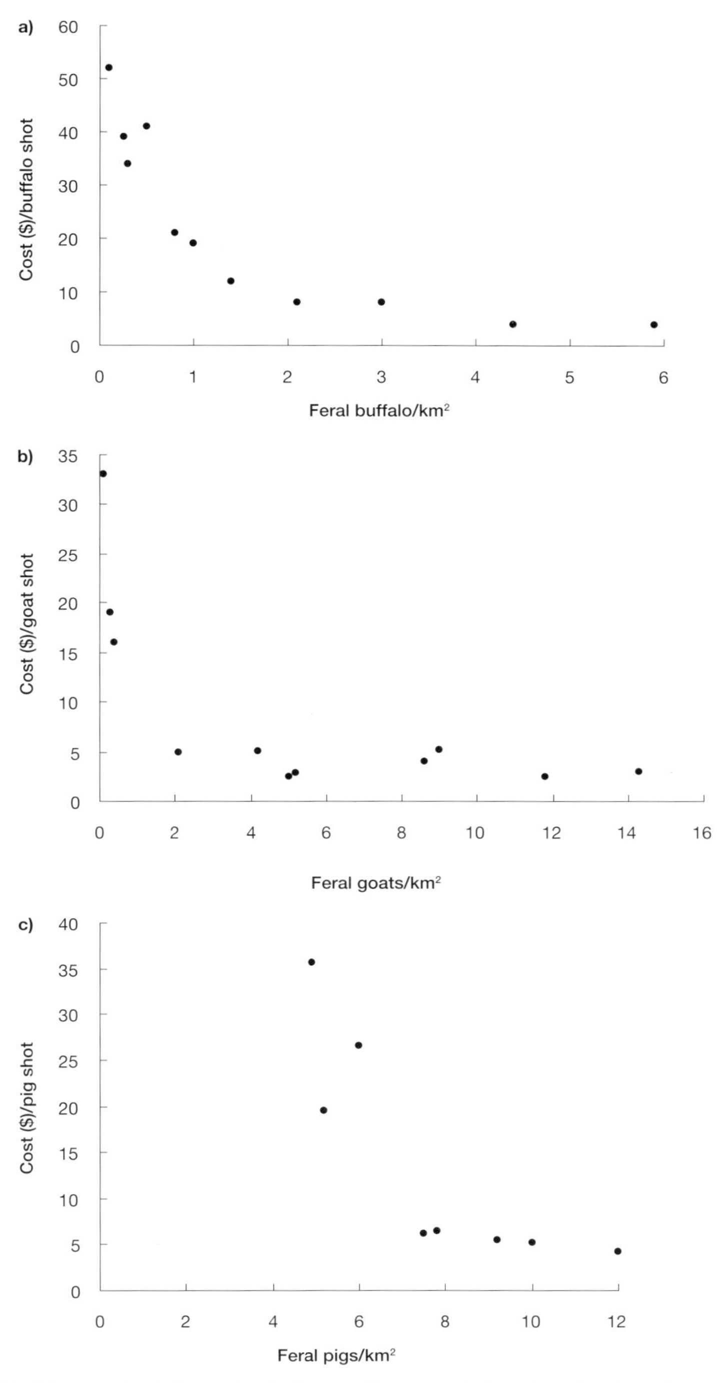

Figure 3.12: Observed relationships between the population density of pests and the cost per animal shot. (a) Feral buffalo (after Bayliss & Yeomans 1989). (b) Feral goats (after Pople *et al.* 1998). (c) Feral pigs (after Choquenot *et al.* 1999).

(Fig. 3.12b) and feral pigs (Fig. 3.12c). A linear negative relationship between pest abundance and capture effort was assumed for white-tailed deer (Nielsen *et al.* 1997). This was recognised as conservative and may underestimate effort required at low abundance. An empirical negative exponential relationship between time per capture and deer density was later reported (Rudolph *et al.* 2000). Each of these two relationships for deer displayed some or all of the framework relationship shown in Figure 1.3d. Note that each example above relates to pest abundance, not damage. Marginal responses also occur in the benefits from control such as illustrated in the framework relationships (Fig. 1.4a–c). However, empirical examples of such benefit relationships are difficult to find in the literature.

These empirical results suggest the **marginal response** principle. This states that each unit reduction in pest damage and abundance is more expensive than the previous unit reduction (Table 3.1).

The marginal response can be estimated by regression analysis of data obtained in observational or experimental studies across a range of pest abundance. The focus is on the slope of the regression line, as it estimates the change in the y variable for a unit change in the x variable, i.e. the marginal response. The regression equation may be derived empirically, such as polynomial, or derived theoretically, such as from the functional response relationship of predator–prey theory (Hone 1990; Choquenot *et al.* 1999). Note that it may be necessary to reduce pest abundance, or density, to very low levels to detect the bend in the curve and hence obtain accurate estimates of costs at low abundance.

Substitution

When more than one option exists for pest control (e.g. shooting and trapping) the profitability will depend on the level of each type of control and their respective costs and effectiveness. A manager may consider substituting some trapping for some shooting. The evaluation of the substitution is assisted by the economic principle of **substitution** (Malcolm *et al.* 1996). This states that one method of damage control should be substituted for a different method if there are higher benefits for a given cost (Table 3.1). Method 1 of damage control should be substituted for method 2 if the benefit per unit cost is higher for method 1 than method 2. In an agricultural context, the benefit may be measured as the monetary value of crop yield or livestock production.

The effects of substitution can be estimated by calculating the costs and benefits of different levels of each type of pest control, such as reported for rabbit control (Williams & Moore 1995). Malcolm *et al.* (1996) give an agricultural example where alternative inputs are fertiliser and water. In wildlife damage control the principle of **substitution** requires knowledge of the framework relationships (Fig. 1.4, p. 9) for each control method, that is, the relationship between biodiversity and effort for trapping, the relationship between biodiversity and effort for shooting, and the relationships between costs and effort for each control method.

Conclusion

This chapter has discussed generalities in the control of wildlife damage. Subsequent chapters examine damage control for particular topics, such as biodiversity conservation (Chapter 4), production (Chapter 5), human and animal health (Chapter 6) and recreation (Chapter 7).

Worked examples

1 Estimation of required level of pest control

This worked example illustrates the **pest removal rate** principle. Data on the population density of Belding's ground squirrel (*Spermophilus beldingi*) in part of California were obtained, in and around alfalfa fields. The estimates were obtained at the same time for 15 years (Table 3.2; data are hypothetical but based on a real study). You should estimate the observed annual rates of increase for each year and the average across years. Use those rates to estimate the proportion of ground squirrels that must be removed each year to stop population growth. Compare the estimated required culls with a theoretical estimate from the rate of increase–bodyweight equation of Sinclair (1996), assuming an average adult bodyweight of 280 g (Burt & Grossenheider 1976).

Solution

The annual finite rates of increase (λ) are estimated as the ratio of successive density estimates. The annual instantaneous rates (r) are estimated as the natural logarithms of the finite rates. The average annual instantaneous rate of increase is estimated from the linear regression of

Table 3.2: Estimates of true density of Belding's ground squirrel in an area of California, associated annual rates of population increase and proportions of squirrels to cull annually to stop population growth

Year	Density (/ha)	Finite rate of increase (λ)	Instantaneous rate of increase (r)	Proportion to cull annually
1	20	–	–	–
2	25	1.25	0.22	0.20
3	32	1.28	0.25	0.22
4	27	0.84	-0.17	0.00
5	33	1.22	0.20	0.18
6	40	1.21	0.19	0.17
7	48	1.20	0.18	0.17
8	25	0.52	-0.65	0.00
9	25	1.00	0.00	0.00
10	39	1.56	0.45	0.36
11	37	0.95	-0.05	0.00
12	46	1.24	0.22	0.19
13	57	1.24	0.22	0.19
14	68	1.19	0.17	0.16
15	79	1.16	0.15	0.14

the natural logarithm of density estimates (y values) over years (x values) (Caughley & Sinclair 1994).

Population density of ground squirrels broadly increased over the study period though there were three short declines in density. The annual finite rate of increase varied, with a minimum of 0.52 and a maximum of 1.56. The annual instantaneous rate of increase varied similarly. The average annual instantaneous rate of increase (r) was 0.07 (+/– 0.01 SE), and the maximum was 0.45. The average (0.07) was significantly different from zero (F = 25.09; df = 1,13; $P < 0.001$). The coefficient of determination (r^2) was 0.66. It is concluded that as the average value of r was greater than zero, population density increased.

The annual proportion of squirrels to cull varied from 0.00 when density declined naturally or stayed constant, to a maximum of 0.36. The mean annual proportion to cull across the 15 years was 0.14. The intrinsic rate of increase (r_m) estimated from Sinclair's (1996) bodyweight equation was 2.05 per year, much higher than the observed maximum (0.45). At that rate of increase (r_m = 2.05) the proportion of ground squirrels to cull annually is 0.87 ($= 1 - (1/e^{2.05})$).

Field data from a detailed demographic study of Belding's ground squirrel in California estimated a net reproductive rate (R) equal to 1.356, corresponding to an increasing population size (Sherman & Morton 1984). The proportional reduction in R needed to stop such population growth is given by $p' = 1 - (1/R) = 1 - (1/1.356) = 0.263$. That is, a reduction of 26.3% in females entering the next generation is needed to stop population growth. Note that p' is similar to, but different from, p. The latter estimates an annual proportion to cull; the former estimates a proportional reduction in female young per female per generation. The two estimates converge when the generation interval is one year.

2 Unintended consequences of control

This worked example illustrates the **response to control** principle (Table 3.1). The likely occurrence of damage can strongly influence whether control occurs. Sometimes damage is expected but has not yet occurred. In such situations the decision to initiate control is made with some uncertainty about whether the damage will occur (the **precautionary** principle). The control actions can have unintended consequences. Review the classic story of Jemima Puddle-duck by Beatrix Potter (1908), that illustrates these points. Describe the damage issue, the control that was initiated in response to the likelihood of damage and the unintended consequences of control.

Solution

The farm dogs initiated fox control because of concern that a fox would eat the duck (Jemima) or her eggs. The dogs began control before any damage to the duck or her eggs occurred – it was pre-emptive control. The dogs chased off the fox and the dogs (not the fox) ate the eggs. The behaviour of the dogs could be described as an unintended consequence of fox control, or as an example of pest replacement.

4
Biodiversity conservation

Introduction

Vertebrates such as rats, cats, goats, pigs and snakes have had negative effects on other species following their introduction in many parts of the world. One example is the decline or disappearance of bird species on a tropical island after introduction of the brown tree snake (Savidge 1987; Rodda *et al.* 1997; Fritts & Rodda 1998). Rats threaten many island birds (Diamond 1985) and have been implicated in the decline of a population of birds, the ancient murrelet (*Synthliboramphus antiquus*), on an island in British Columbia (Bertram 1995; Hobson *et al.* 1999). Mongoose (*Herpestes auropunctatus*) have reduced abundance of lizards on islands in the Pacific Ocean (Case & Bolger 1991). Exotic herbivores and carnivores have modified the vegetation and bird communities on subantarctic islands (Chapuis *et al.* 1994). The proportion of bird species going extinct on oceanic islands was positively related to the number of exotic predatory mammals that established on the islands after European colonisation (Blackburn *et al.* 2004, 2005).

Vertebrates can also cause problems in continental areas. Feral pigs have adversely affected the vegetation of conservation reserves, such as in Great Smoky Mountains National Park (Bratton 1975), and the vegetation (Hone & Stone 1989; Stone & Stone 1989) and invertebrates (Vtorov 1993) in Hawaii Volcanoes National Park. Feral horses have affected native vegetation in Cumberland Island National Seashore (Turner 1988) and salt marshes in the Shackleford Banks, North Carolina (Levin *et al.* 2002) in the US, and stream banks and vegetation in Kosciuszko National Park in Australia (Dyring 1990). Brushtail possums and black rats are important predators of native birds, such as kokako (*Callaeas cinerea wilsoni*) in New Zealand (Innes *et al.* 1999). Mink (*Mustela vison*) affect waterbird abundance in southern England (Ferreras & Macdonald 1999) and hedgehogs (*Erinaceus europaeus*) are predators of eggs of wading birds in the Western Isles, Scotland (Jackson 2001).

Introduced species are not the only vertebrates that affect biodiversity. Native species can have undesirable effects on other native species, which may require conservation efforts by management. For example, white-tailed deer cause changes in vegetation in the US, and kangaroos affect shrub vegetation in national parks in southern Australia. Many such examples were reviewed by Goodrich and Buskirk (1995).

Vertebrate pests may have large effects on other species which are the focus of conservation efforts. However, there are many possible causes of species and population decline, such as habitat clearing and change in fire regimes (see Burbidge & McKenzie 1989). Caughley (1994) included vertebrates as pests in the 'evil quartet' of processes that have caused plant and animal species worldwide to become threatened or endangered.

This chapter examines principles pertaining to the conservation of biodiversity. 'Biodiversity' is used to describe the variety of life; it can refer to variety at the community, population or genetic level. An interesting and useful historical discussion of the term is given by Harper and Hawksworth (1994). Economic analyses of biodiversity conservation are not examined in detail in this chapter. There are published examples however. Benefit/cost ratios of predator management have been estimated for conservation of marine turtles in Florida (Engeman *et al.* 2002) and Puerto Rica parrots (*Amazonia rittata*) (Engeman *et al.* 2003). The cost effectiveness of managing kokako at different sites in New Zealand has been examined (Cullen *et al.* 2005). Details of various economic analyses are discussed in Chapter 5.

There are six principles relating to conservation of biodiversity (Table 4.1). The principles are grouped into those concerned with conservation of species and those concerned with conservation of communities. The generic framework principles illustrated in Figures 1.3 and 1.4 (pp. 8–9)

Table 4.1: Principles of damage control for conservation of biodiversity at the species and community levels of organisation

Conservation of species
4.1: Population limitation. Increasing abundance of a threatened species involves identifying and manipulating the factor(s) limiting abundance and rate of increase, such as another species (the pest)
4.2: Reintroductions. If an interacting pest species (competitor, predator, pathogen or parasite) causes a species to become extinct, reintroduction of the species should occur only after the effects of the pest species have been reduced
4.3: Threshold population parameters. To increase abundance of a species being conserved, key population parameters must have values higher than critical threshold values
Conservation of communities
4.4: Threshold habitats. There is a threshold proportion of linked habitats, suitable for a species being conserved, in which a co-existing pest must be controlled
4.5: Community effects. Reducing pest abundance causes changes in richness and other parameters of species being conserved
4.6: Multiple pests. The control of one pest species may affect a community differently from the control of another pest species

are applied here with the response variables being population parameters, such as density, and community parameters, such as species richness.

Conservation of species

When conserving a species, it is useful to know what affects its distribution and abundance. Three principles are relevant to this discussion.

Population limitation

A predator may limit the abundance of a rare prey species. If the prey species is the focus of conservation efforts, plans must be developed to manipulate predator abundance. The predator, native or exotic, is deemed to be a pest. The key planning feature is the identification of the predator and its role in limiting prey abundance. If predator control starts without identifying whether the predator in fact limits prey abundance, the prey species may show no response. The same holds true if the second species is a competitor, a parasite, a pathogen or a herbivore, and hence not a predator. For example, evaluating the role of brown tree snakes in limiting abundance of birds on Guam involved carefully identifying and evaluating alternative hypotheses (Savidge 1987; Rodda *et al.* 1997; Fritts & Rodda 1998).

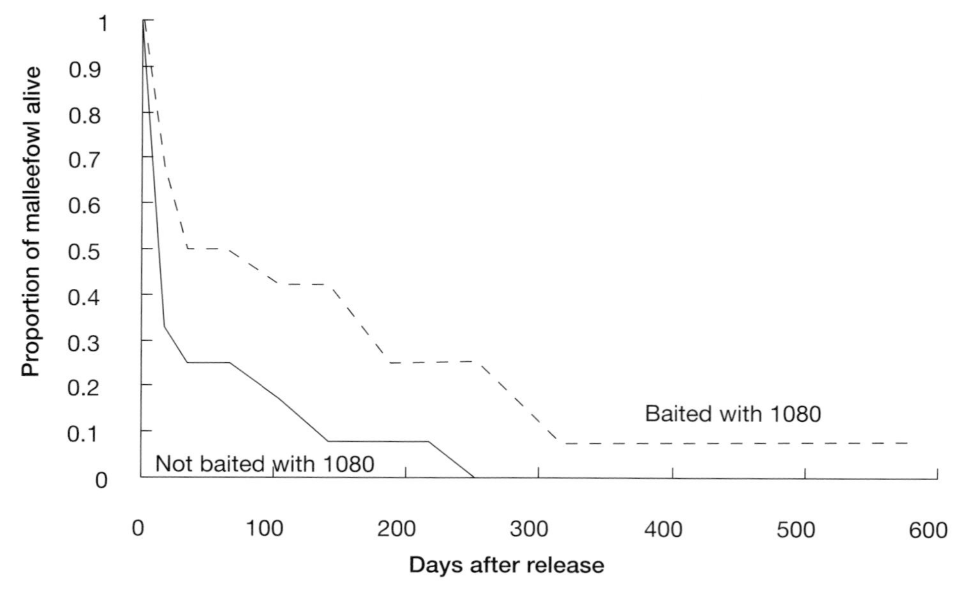

Figure 4.1: Trends in survival of malleefowl released into an area where fox baiting with sodium monofluoroacetate (compound 1080) was occurring (dashed line) and an area where fox baiting had not occurred (solid line).

Source: Data from Priddel and Wheeler (1997; Table 2).

The abundance of the Lord Howe Island woodhen (*Tricholimnas sylvestris*) was very low until feral pigs, inferred as limiting woodhen abundance by predation, were removed (Miller & Mullette 1985). Red fox predation is believed to be an important cause of malleefowl (*Leipoa ocellata*) mortality in southern Australia and fox baiting significantly increased survival of released birds (Fig. 4.1) (Priddel & Wheeler 1997). However, predation by foxes was not the only limiting factor. Food was also a limiting factor (Priddel & Wheeler 1990). The foliage cover of fuchsia (*Fuchsia excorticata*) in parts of New Zealand was significantly lower where possum abundance was higher (Fig. 4.2) (Pekelharing *et al.* 1998). The abundance of rock-wallabies (*Petrogale lateralis*) in parts of Western Australia increased (Fig. 4.3) at sites where baiting occurred for red foxes (Kinnear *et al.* 1998). In parts of New Zealand, kokako increased in abundance after identification of predators such as brushtail possums and black rats (Fig. 4.4) and their control (Innes *et al.* 1999). The sighting rate of numbats (*Myrmecobius fasciatus*) in Dryandra woodland in Western Australia increased after baiting against red foxes. However, after several years of ongoing baiting the sighting rate decreased to pre-baiting levels

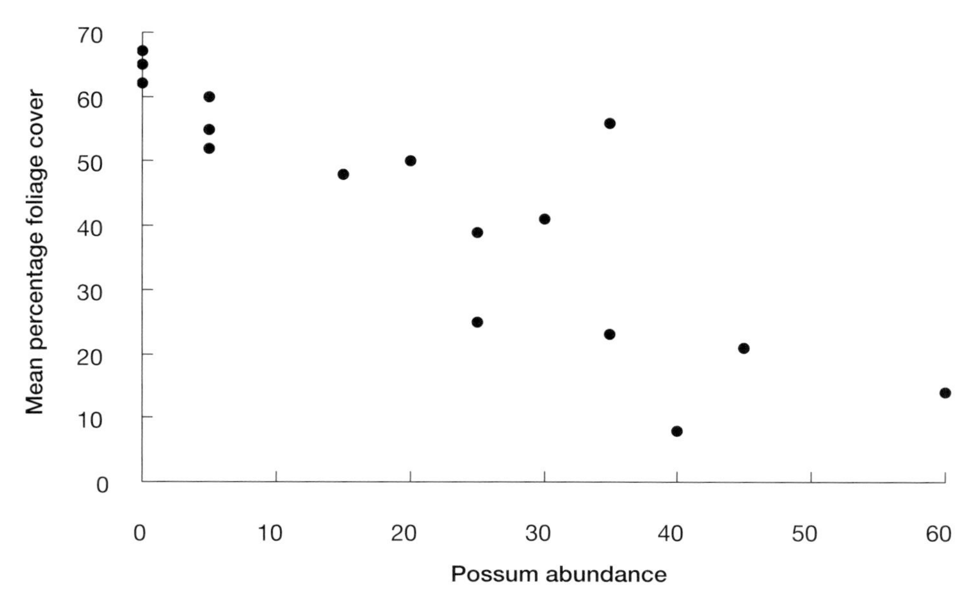

Figure 4.2: The relationship between mean foliage cover of fuchsia and abundance of brushtail possums. The latter was assessed as the possum trap catch rate at five sites over 4–5 years.

Source: Modified from Pekelharing *et al.* (1998, Fig. 3).

(Friend & Thomas 2003). A similar increase and then decrease after fox baiting was reported for a translocated chuditch (*Dasyurus geoffroii*) population in Western Australia (Morris *et al.* 2003). The increases and subsequent decreases may be examples of eruptive fluctuations of wildlife populations similar to those reported elsewhere (Caughley 1970; Forsyth & Caley 2006). The increased abundance may be related to reduction in predation and the decreased abundance to food limitation.

A model of the relative responses of many plant species to differing abundance of red deer (*Cervus elaphus*) and brushtail possums in New Zealand was proposed by Nugent *et al.* (2001). Some plant species increased and others decreased as herbivore abundance increased. Such empirical patterns illustrate the framework relationship for biodiversity shown in Figure 1.3a (p. 8) – biodiversity varies with pest abundance.

A generalisation suggested by the above results is that to increase the abundance of a threatened species involves identifying and manipulating the factor(s) limiting abundance and rate of increase, such as another species (the pest). This is the **population limitation** principle (Table 4.1).

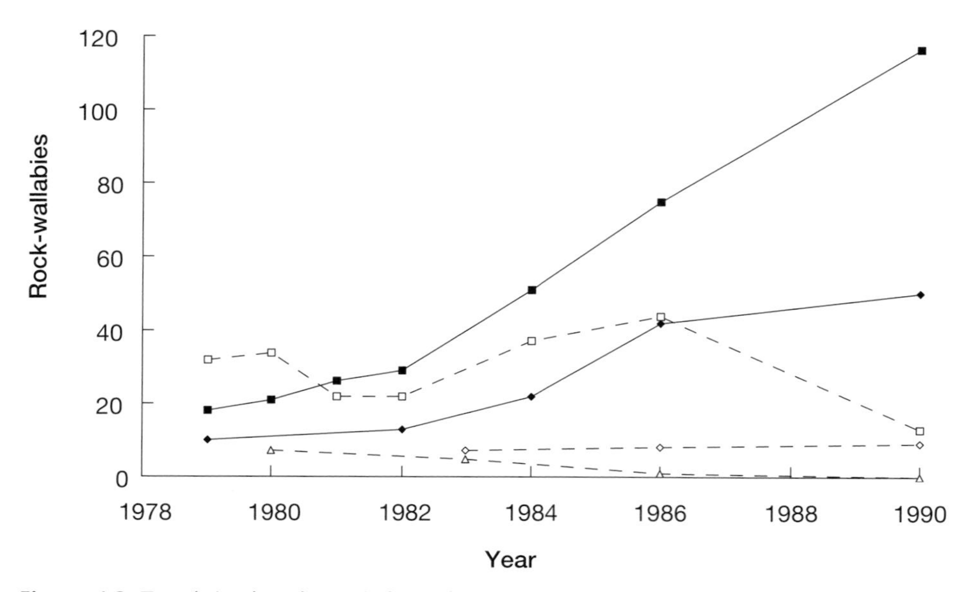

Figure 4.3: Trends in abundance (adjusted census estimates) of rock-wallabies over time at two sites where baiting for foxes with compound 1080 occurred (solid lines and symbols) and at three sites where no baiting occurred (dashed lines and open symbols).

Source: Data from Kinnear *et al.* (1998, Table 1).

In terms of population dynamics, the change in abundance over time (population growth rate) is determined by demographic rates. These in turn are determined by trophic factors such as food, predators, competitors and pathogens, and factors internal to the population, such as genotype and behaviour (Sibly & Hone 2002). Experiments can predict population responses to manipulation of demographic rates and trophic factors. Prediction has also been done by many types of modelling, as described by Bradbury *et al.* (2001).

Pest control, such as predator control, may not cause a change in prey abundance even when the predator eats the prey. The predation may simply replace a different source of mortality. That occurred when there was no change in abundance of native bush rats (*Rattus fuscipes*) after control of red fox, in southern Australia (Banks 1999). The lower mortality from predation was compensated for by an increase in other causes.

Predator removal to protect bird populations was evaluated in a meta-analysis by Cote and Sutherland (1997). The mean effect size of predator control was not significantly different from zero, when assessed as a change in bird breeding population size. In contrast, there were significant effects on hatching success and post-breeding population size.

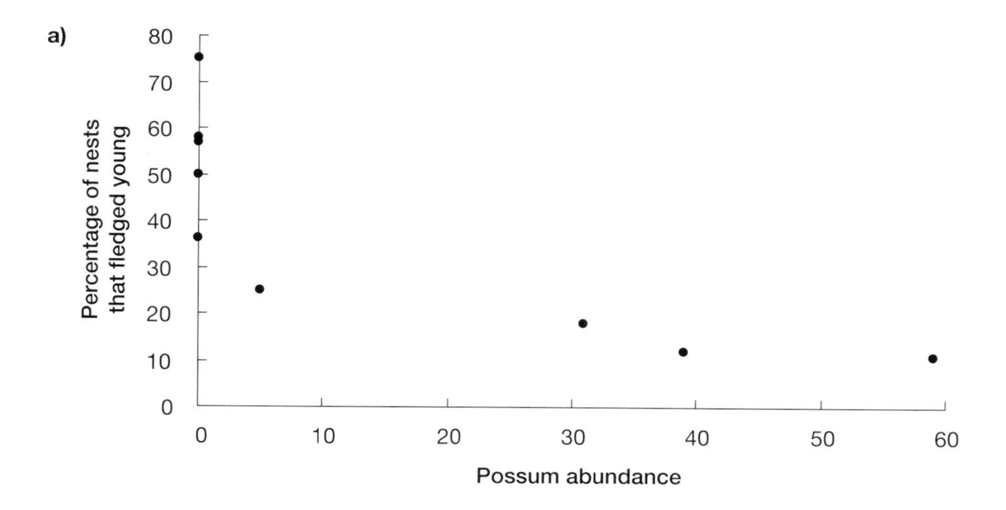

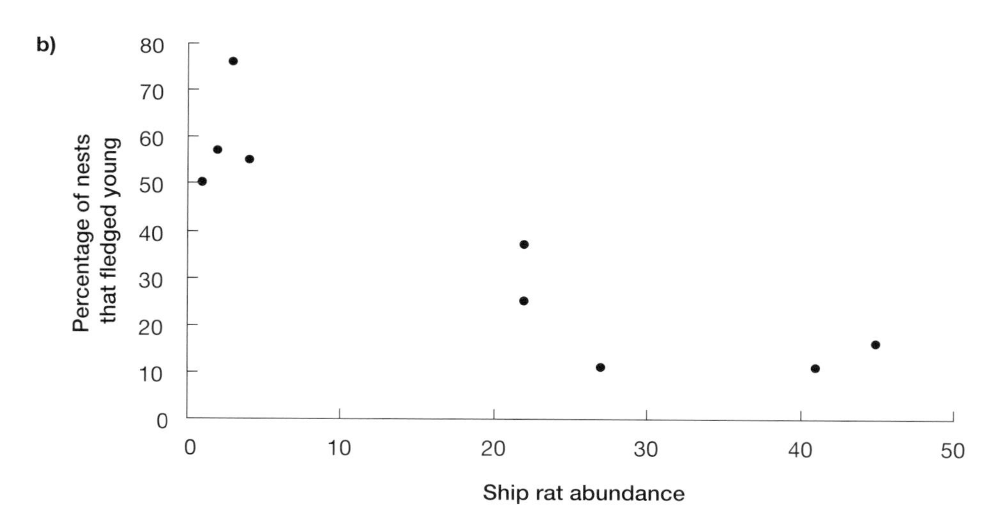

Figure 4.4: Relationships between abundance of (a) brushtail possums and (b) ship rats, and nest success of kokako, as assessed by the percentage of nests that fledged young.

Source: Redrawn from Innes *et al.* (1999, Fig. 7).

Control of red fox was associated with increased abundance of California clapper rail (*Rallus longirostris obsoletus*) (Harding *et al.* 2001).

Reintroductions

A conservation strategy used in several parts of the world is to reintroduce a species into parts of its former range. The Hawaiian goose, the nene (*Nesochen sandvicensis*) can be readily bred in captivity and released into the wild, though the wild populations may not be viable without repeated reintroductions (Stone 1989). However, in Australia, Short *et al.* (1992) showed that many reintroductions of threatened species, such as wallabies, were unsuccessful. They postulated that this occurred because the threatening processes were not removed or reduced in intensity prior to reintroductions. The threatening process may be other wildlife, such as predators or herbivores. Reintroductions of numbats (Friend & Thomas 2003) and chuditch (Morris *et al.* 2003) in parts of Western Australia occurred after extensive baiting against red foxes, a reputed predator of the two species.

If an interacting pest species (competitor, predator, pathogen or parasite) causes a species to become extinct, reintroduction of the species should occur only after the effects of the pest species have been reduced. This is the **reintroductions** principle (Table 4.1). Reducing the effects may involve reducing pest abundance, even to extinction. Reintroductions are the opposite of eradications. Hence this principle is linked to the **eradication conditions** principle (Table 3.1, p. 39).

Threshold population parameters

For a species to increase in abundance or population density the net reproductive rate (R) must be greater than 1.0. Net reproductive rate (Krebs 2001) is the sum across age classes of the average number of female young per female per generation ($R = \Sigma\, l_x m_x$) (Fig. 3.8a–d, p. 54). Alternatively, if individuals in a species being conserved are eaten by a predator before they reach average reproduction age, abundance of the species will decline. The generation interval (T) is the average age of mothers giving birth (Caughley 1980). A population declines when, for example, the average age at which females get eaten by a predator is less than the generation interval. The analogous argument with respect to a lethal or sterilising pathogen would be that for a population to increase, the generation interval (T) should be less than the average age of first infection (A) with the pathogen.

A related topic is whether abundance needs to increase above a minimum viable population (MVP). The MVP is the population that will ensure, at some acceptable level of risk (often 95%), that the population will persist for a specified time (often 100 years) (Gilpin & Soule 1986). There are no hard and fast rules about how many is enough, only a generalisation that the risk of extinction is usually greater for smaller than for larger populations. For example, the value of 50 animals is derived as the solution to the equation for loss of genetic variance (loss = $1/(2N_e)$), where the rate of loss is 1% per generation (0.01) and N_e is effective population size (Caughley & Sinclair 1994). Krebs (2001) considered that the smaller the population the greater the risk that chance events can lead to extinction was a general principle in conservation biology. The concept of MVP can be interpreted in mathematical models of rate of increase and population density, as the lower density, k, at which the instantaneous rate of increase equals zero ($r = 0$) (Fig. 3.5, p. 48).

Some empirical studies support the concept of a threshold population size. Evaluation of introductions of ungulates to New Zealand (exotic not native) reported an apparent threshold of six individuals. In the 13 species studied, the introduction was successful if more than six individuals were introduced (Forsyth & Duncan 2001). A study of reintroductions of nine ungulate species in North America, Europe and the Middle East reported a similar threshold of eight individuals (Komers & Curman 2000).

To increase abundance of a species being conserved, key population parameters must have values higher than critical threshold values. This is the **threshold population parameters** principle (Table 4.1). The principle is closely linked to the **pest removal rate** principle (Table 3.1, p. 39), though our focus is on the species being threatened; the **pest removal rate** principle focuses on the pest.

For key population parameters to have values higher than threshold values, there may need to be replicated linked populations that include the range of genetic variation. The parameter values of interest are average values across the replicates. The replicate populations may be less vulnerable to extinction than one population, and linkages between populations can facilitate movement and genetic exchange. The amount of movement required to reduce the extinction risk has been variously estimated as one individual per generation to 3–10 individuals per generation (Vucetich & Waite 2000). As vertebrate pests may limit movement, they have implications for genetic diversity in a species being conserved. A key message is that genetic variation may not guarantee population survival; it is the effect of genetic variation on key population parameters that is functionally important and determines the fate of the population.

Conservation of communities

Species occur in a community of many species. It is useful to understand the role of pests and their effects when attempting to conserve a community. Three principles are involved in the conservation of communities and wildlife damage control.

Threshold habitats

If the proportion of suitable habitats occupied by a species drops below a threshold level, the species will decline to extinction. This principle was derived mathematically for a metapopulation (Levins 1969; Lande 1987, 1991; Bascompte & Rodriguez-Trelles 1998) and has since been reviewed extensively, for example by Begon *et al.* (1996) and Caughley and Gunn (1996). The concept was applied to the habitat of northern spotted owl (*Strix occidentalis caurina*) in North America (Lande 1991). The concept implies that pests, such as predators or competitors, that threaten the species being conserved, need to be controlled in at least the same proportion of suitable habitats. The strategy of enclosing feral pigs in small areas in Hawaii Volcanoes National Park (Hone & Stone 1989; Katahira *et al.* 1993) is an attempt to increase their extinction rate and decrease their colonisation rate by creating a set of isolated, not linked, populations. If the fences 'leak', a metapopulation may be created.

These theoretical considerations suggest a generalisation, the **threshold habitat** principle (Table 4.1). This principle states that there is a threshold proportion of linked habitats, suitable for a species being conserved, in which a co-existing pest must be controlled.

Community effects

Vertebrate pests may act as agents of disturbance by physical activities, such as scratching, digging and burrowing, and change species richness or diversity of plants and animals. The intermediate disturbance hypothesis (Grime 1973; Connell 1978) states that species richness or diversity increases as the level of disturbance increases, but that at high levels of disturbance richness or diversity declines. For example, species diversity of a prairie grassland in Arizona was highest at intermediate levels of disturbance (Fig. 4.5) by pocket gophers (*Thomomys bottae*) (Martinsen *et al.* 1990). Note that this is not a specific example of a vertebrate pest issue, but an example of the disturbance–diversity issue. The disturbance–diversity relationship is not always strong and peaked, however (Mackey & Currie 2001). The intermediate disturbance hypothesis predicts that as pest abundance is reduced species richness will initially increase, then decrease. It is necessary to identify the management aim – is it to increase species richness, to maintain it or to decrease it?

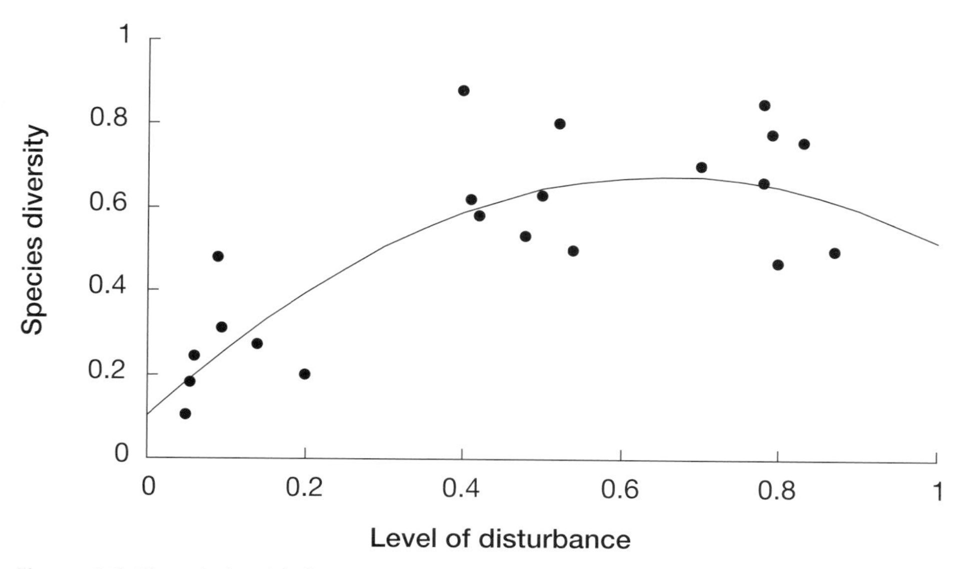

Figure 4.5: The relationship between species diversity of a prairie grassland in Arizona and the level of disturbance by pocket gophers.

Source: Redrawn from Martinsen *et al.* (1990, Fig. 2).

In community ecology, the predation hypothesis (Paine 1966) says that predation is a strong determinant of species richness in an area. Reducing predation can decrease species richness by increasing interspecific competition between prey species. The effect occurs when predation changes the competitive interactions between prey species: if the predator preys on a species which otherwise outcompetes other prey species then the predation allows extra prey species to coexist in the community. This is also called the predator-mediated coexistence hypothesis (Caswell 1978). An example of the principle involved excluding rabbits from one small plot on Macquarie Island, in the Southern Ocean. The number of plant species in that plot declined from 10 to three over 16 years (Fig. 4.6) (Copson & Whinam 1998). Unfortunately, no experimental controls were monitored equally intensively, though photographic evidence suggests such changes did not occur outside the exclosure.

A decline in species richness occurred on an island off Mexico after exclusion of rabbits; however, richness also declined on a nearby island where rabbits occurred (Donlan *et al.* 2002). Another example was the increase in abundance of endemic springtails (soil invertebrates) in a Hawaiian rainforest (Fig. 4.7) after removal of feral pigs (Vtorov 1993). Cosmopolitan species of springtails declined over the same time period. The predation hypothesis is different from the disturbance hypothesis

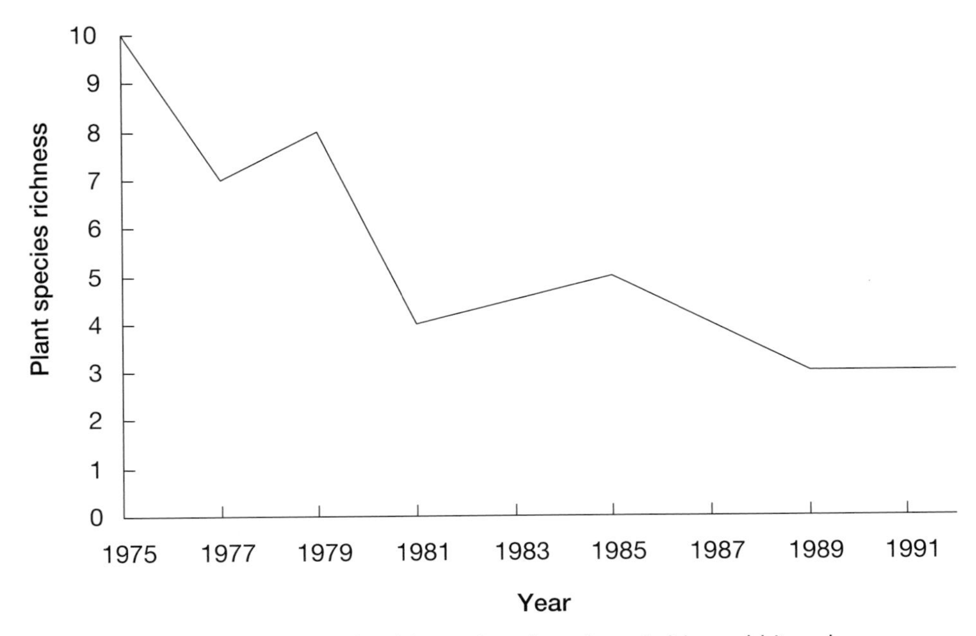

Figure 4.6: Trends in plant species richness (number of species) in a rabbit exclosure on Macquarie Island.

Source: Data from Copson and Whinam (1998, Table 5).

when the pest species is a predator. Predation and disturbance are two of the eight hypotheses used to account for biodiversity gradients in nature (Krebs 2001).

The grazer-reversal hypothesis (Proulx & Mazumder 1998) suggests that plant species richness decreases with high grazing pressure in nutrient-poor ecosystems, while richness increases with high grazing in nutrient-rich ecosystems. This hypothesis suggests that opposite responses to pest control by reduction of grazing could be expected in nutrient-poor (response is an increase) and rich (response is a decrease) ecosystems. Hence the framework relationships of Figure 1.4 (p. 9) show a variety of patterns between biodiversity and levels of pest control. The difference between the predation and grazer-reversal hypotheses is like the difference between a main effect and an interaction in analysis of variance – the former hypothesis is more general and the latter is more specific.

In addition to changes in the number of species there may be changes in relative abundance of species. That is, the frequency distribution of abundance across all species may change, to be dominated by fewer species. Such changes apparently occurred on Macquarie Island, a

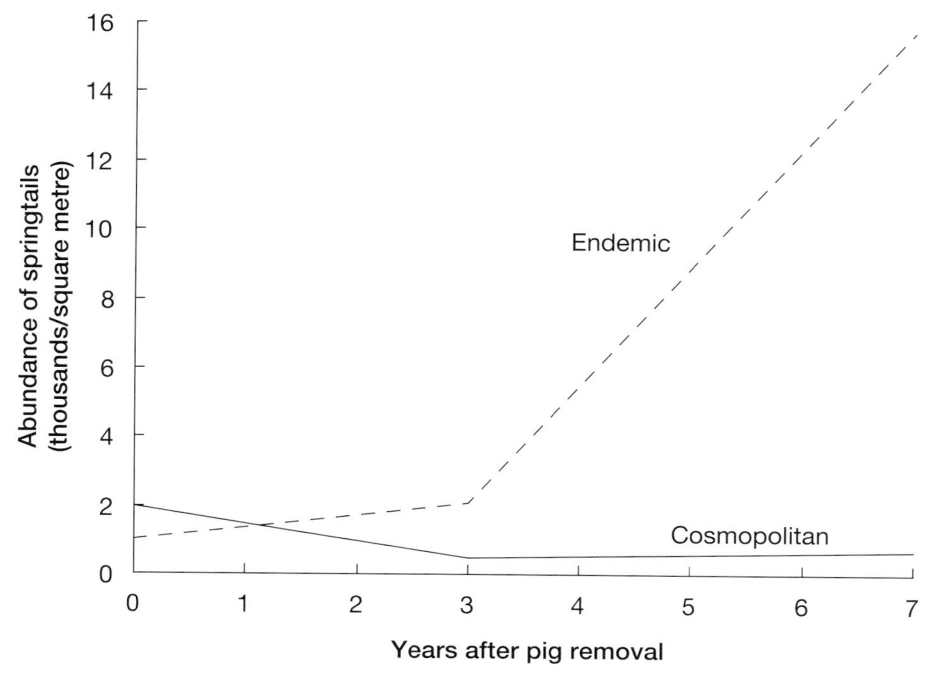

Figure 4.7: Trends in abundance of soil invertebrates (springtails) since feral pigs were excluded in Hawaii Volcanoes National Park. Cosmopolitan species (occurring naturally in many areas of the world, including Hawaii) are shown by the solid line and endemic species (occurring naturally only in Hawaii) by the dashed line.

Source: Modified from Vtorov (1993, Fig. 3), with Vtorov's time period of 2–4 years shown here as 3 years.

subantarctic island, after exclosure of rabbits (Copson & Whinam 1998). An analogous change occurred in a grassland in England after prolonged fertiliser application (Begon *et al.* 1996). Different tree species showed different responses to culling of brushtail possums in New Zealand (Payton *et al.* 1997), including some that showed no response.

Changes may also occur in other properties of biological communities. One trophic structure hypothesis (the world is white, yellow and green: Oksanen *et al.* 1981; Pimm 1991) states that the relative number of species in each trophic level varies in environments of different productivities. Predators are rarer in low-productivity environments than in high-productivity environments. Pimm (1991), Oksanen and Oksanen (2000) and Krebs (2001) offer interesting discussions of this and other relevant hypotheses.

The change in species composition and abundance of species over time is called succession, and many hypotheses have been developed to explain it (Begon *et al.* 1996; Krebs 2001). The inhibition hypothesis of succession (Begon *et al.* 1996) predicts that removal of or reduction in a key species will lead to changes in a community; species and communities change according to the factors that inhibit them. Control of a pest species may alter the successional changes, so the community goes in a different direction from the one it would have taken with no pest control.

Reducing pest abundance causes changes in richness and other parameters of species being conserved. This is the **community effects** principle (Table 4.1), a community-level version of the **population limitation** principle.

Multiple pests

In some communities there are many pest species. For example, in the forests of Hawaii there are feral pigs, mongoose and rats. Different pest species may have different roles in a community. One pest may compete with a threatened species, another pest species may be a predator. Reducing the abundance of the competitor will cause different effects from reducing the abundance of the predator, as described in general terms by Pimm (1987). If there are several pest species that are competitors or several that are predators, the effects of each competitor or predator on the threatened species are probably different. Some species have a greater role in a community than others. In community ecology these are called keystone species (Krebs 2001) – the interactions between such species are stronger. Mathematical modelling, used to study this topic, has demonstrated the potentially different effects of controlling rabbits or cats (Courchamp *et al.* 1999b) or controlling rats and cats (Courchamp *et al.* 1999c) on oceanic islands, while trying to conserve birds.

A species' response to control of a different species will be influenced by the niche breadth of each species. If niches overlap the second species may expand its realised niche and partially replace the first (controlled) species. Such interactions were used to argue for multi-species management of sympatric exotic carnivores in southern Australia (Molsher 1999) and sympatric exotic herbivores in the alps of New Zealand (Forsyth *et al.* 2000).

The **multiple pests** principle states that the control of one pest species may affect a community differently from the control of another pest species (Table 4.1). The pest species that is controlled or removed may simply be replaced by a different pest species (the **response to control** principle; Table 3.1, p. 39).

Conclusion

This chapter has described and examined six principles of the conservation of biodiversity. The application of the framework illustrated in Figures 1.3 and 1.4 (pp. 8–9) was also described. The next chapter examines principles in wildlife damage control for production, such as in agriculture, forestry and fisheries.

Worked examples

1 Controlling impacts and a pest population: a quantitative example

This example demonstrates how to quantitatively link aspects of the impacts of a pest, pest population dynamics and pest control, studying feral pigs in a national park. The approach is simplified but captures the key elements. The example illustrates the **damage extent** principle (Table 2.1, p. 16) and the **eradication conditions** and **pest removal rate** principles (Table 3.1, p. 39), and is linked to the **community effects** principle (Table 4.1, p. 69). The last occurs through the effects of ground rooting on plant species richness (Table 2.4 and Fig. 2.6, pp. 30–31).

Assume that feral pigs are having conservation impacts by their rooting up of the ground, as illustrated in Chapter 2. A relationship has been derived empirically between the percentage of ground rooted by pigs and an index of pig abundance (Hone 2002). The equation is:

$$\%GR = 1.21(\%D)^{0.408} \quad (4.1)$$

where $\%GR$ is the percentage of ground rooted at the time of measurement (hence represents a cumulative measure) and $\%D$ is the percentage of plots with fresh pig dung (a linear index of pig abundance). The relationship described in equation 4.1 is positive and curved (concave down), rising quickly at first then rising more slowly, as the exponent (0.408) is less than 1.0 but greater than 0. The index of pig abundance is positively correlated with observed pig abundance, the equation for which is:

$$\%D = 5.15 + 2.97(N_{t+1}) \quad (4.2)$$

where N_{t+1} equals observed abundance of feral pigs at time $t+1$ during intensive area counts (Hone 1995).

Assume the population dynamics are described by logistic growth, a simple form of density dependence, an equation for which is:

$$N_{t+1} = N_t + r_m N_t(1 - (N_t/K)) \quad (4.3)$$

where N_{t+1} = abundance at time $t + 1$, N_t = abundance at time t, r_m = the annual intrinsic rate of increase, and K = carrying capacity (= N_t when $r = 0$ and $\lambda = 1.0$). For feral pigs, assume r_m = 0.742/year and K = 1000 pigs. These data are based on research results in Namadgi National Park in south-eastern Australia (Hone 2002). The time step in equation 4.3 is one year.

Pig control can be described by modifying equation 4.3 with the inclusion of an extra term, H, for the number of pigs removed in annual control. Equation 4.3 becomes:

$$N_{t+1} = N_t + r_m N_t(1 - (N_t/K)) - H \quad (4.4)$$

Estimate the effects on impacts and pig abundance of removing the following number of pigs during annual control: 0, 100, 185.5, 200 and 250 pigs. You do not need to specify the type of control, simply assume the pigs are removed. Assume that at the start N_t = 1000 pigs = K.

This example is simplistic. Identify and describe some limitations of the approach. Discuss the implications of such limitations and suggest how each may be overcome in a more realistic example.

Solution

Key steps in this example involve developing an equation for obtaining the estimates. For impacts, this involves substituting equation 4.2 into equation 4.1. This gives:

$$\%GR = 1.21(5.15 + 2.97N_{t+1})^{0.408} \quad (4.5)$$

which is shown in Figure 4.8, and substituting equation 4.4 into equation 4.5 to obtain:

$$\%GR = 1.21(5.15 + 2.97[N_t + 0.742N_t(1 - (N_t/1000)) - H])^{0.408} \quad (4.6)$$

This equation can be solved for each value of H. To estimate the effects of control on pig abundance, use equation 4.4.

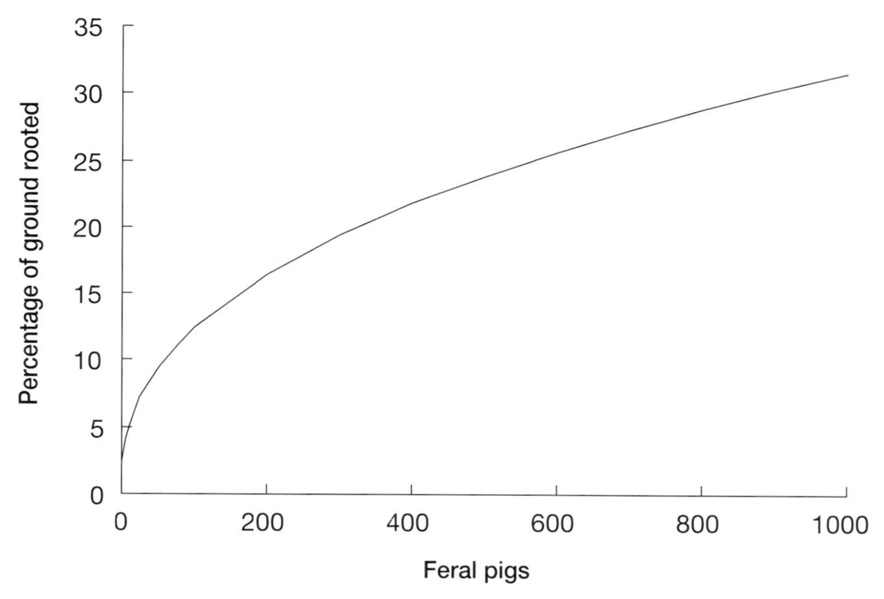

Figure 4.8: The empirical relationship between the percentage of ground rooted by feral pigs and feral pig abundance in south-eastern Australia.

Source: After Hone (2002).

When no feral pigs are controlled ($H = 0$), ground rooting is constant (Fig. 4.9a). As H increases there is a small reduction in ground rooting. At the highest level of control ($H = 250$) there is a slow then rapid reduction in ground rooting. The rooting is reduced to zero in year 10 when $H = 250$ pigs are removed per year. These trends are associated with a slow reduction in pig abundance with low levels of control ($H = 100$) and a faster reduction once control exceeds 185.5 pigs annually (Fig. 4.9b). Pigs are removed completely in year 9 when $H = 250$ pigs. The level of annual removals of 185.5 pigs was included as this is the maximum sustained yield (MSY), as MSY = $r_{\mathrm{m}}K/4 = 0.742 \times 1000/4 = 185.5$ pigs. Once annual removals exceed MSY, pig abundance declines to zero. In Figures 4.9a and b there is a change in the shape of the curves for $H = 200$ and $H = 250$, when $N_t < 500$. Why? The MSY occurs at $N_t = K/2 = 500$ pigs so once abundance is less than 500 sustained yield declines. However, as annual control in this example stays the same then abundance declines as control is greater than sustained yield.

The example is simplistic, though it uses real data and results. Possible improvements include the topics of impacts, dynamics and control.

Alternative measures of impacts could be abundance of some threatened species, and a measure of a community parameter such as plant

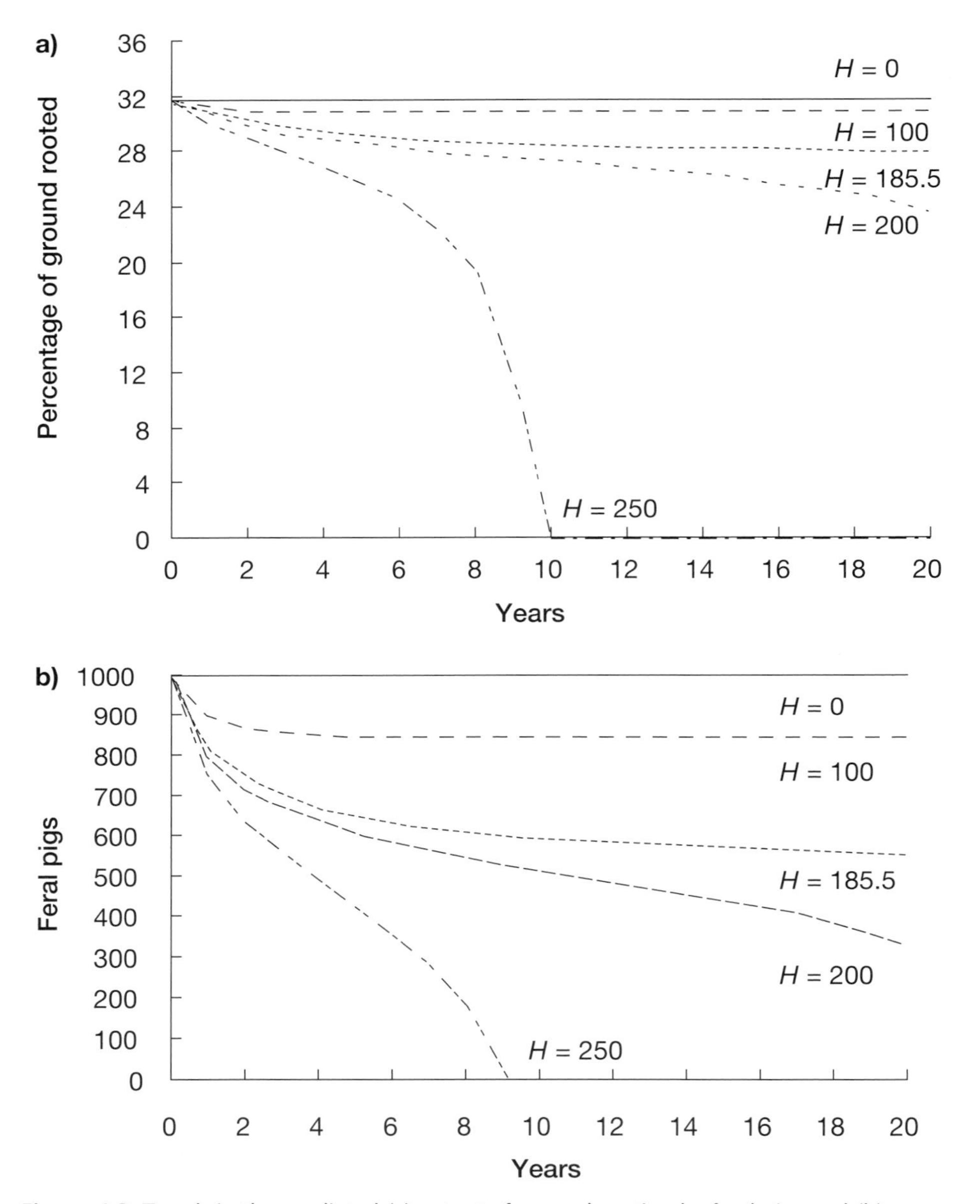

Figure 4.9: Trends in the predicted (a) extent of ground rooting by feral pigs and (b) abundance of feral pigs, based on equations 4.6 and 4.4 respectively. The effects of differing numbers (*H*) of feral pigs removed annually by control are also shown.

species richness. The latter was reported by Hone (2002). Equation 4.2, linking the extent of ground rooting and pig abundance, may be biased as it suggests that even if there are no pigs there is some ground rooting (the intercept on the y axis is 5.15, not 0). Further research would be needed to assess whether the intercept is different from 0.0, or whether it is an example of the issue illustrated in Figure 2.2c (p. 21) of a bias in the method, possibly due to ground disturbance by a different species.

For dynamics, alternative approaches could be to assume the intrinsic rate of increase (r_m) is a variable rather than a constant. That approach was used to examine harvesting issues (Beddington & May 1977). Another possibility could be to assume pig abundance is related to food availability using a numerical response, as reported by Choquenot (1998).

For control, an alternative is to assume that annual removals decrease as abundance decreases, as pigs become harder to find and costlier to remove (Hone 1990; Choquenot *et al.* 1999) (see also Figures 3.10 and 3.12, pp. 59 and 61).

2 Evaluation of predator control

A carnivore species is declining and concern has been expressed about its long-term survival. The decline is believed to be caused by interspecific competition for food by another carnivore, the red fox. Fox baiting, with sodium monofluoroacetate (compound 1080), occurred over several years and the threatened carnivore population increased rapidly. Interpret the results and suggest the mechanism for the observed trends. Describe the data needed to support the explanation, and other possible explanations. This example illustrates the **population limitation** principle (Table 4.1).

Solution

The simple interpretation is that fox baiting reduced fox abundance and so reduced interspecific competition, allowing increased abundance of the threatened carnivores. An alternative explanation is that the baits used to poison foxes were highly acceptable to the native (threatened) carnivore, which also ate the baits. Some wildlife species are very tolerant of 1080 (King *et al.* 1989); if the carnivore were such a species then few individuals would die. Or perhaps the carnivore population was food-limited prior to the baiting, and the fox baiting was essentially a food supplementation study – the carnivore abundance increased because of the extra food.

There may or may not have been a change in fox abundance as a result of fox baiting. For example, a substantial change (reduction) in fox abundance occurred after fox control in one study (Banks *et al.* 1998).

However, studies have also reported a lack of significant effects of fox baiting on fox abundance (Molsher 1999; Greentree *et al.* 2000). Another alternative explanation is that the fox and carnivore shared a common parasite that limited the abundance of the carnivores. Such an event would show apparent competition of the type discussed by Krebs (2001). When baiting reduced fox abundance, parasite prevalence decreased and the abundance of native carnivores increased.

Key data are required to discriminate between alternative explanations. The initial competition hypothesis requires data on the resource being competed for and a demonstration that it was in short supply prior to fox baiting, that fox abundance dropped because of baiting and that abundance of carnivores increased because of extra availability of the resource. The food supplementation hypothesis requires data on the carnivore's tolerance of 1080, its acceptability and uptake of baits, the simultaneous occurrence of the same population response on sites where unpoisoned baits were distributed, and a lack of population response on sites where no baits were distributed. The shared parasite hypothesis requires identification of a shared parasite, demonstration of its pathogenicity in the carnivores and its decline in prevalence when fox abundance declined after fox baiting. More than one hypothesis could be correct.

5
Production

Introduction

Human activities such as agriculture, forestry and fisheries aim to produce food and/or fibre. Vertebrate pests can reduce crop, livestock, forestry and fisheries yields, as described in Chapter 2. Some studies have focused on assessing the type and level of damage, such as badger damage in England and Wales to fences and crops and by predation (Moore *et al.* 1999), wildlife damage to crops in general (Leopold 1933; Conover & Decker 1991) and to corn in particular (Wywialowski 1996) in parts of the US, and crop damage in parts of Uganda (Naughton-Treves 1998; Naughton-Treves *et al.* 1998). Some studies have focused on specific pest species, such as Belding's ground squirrel and damage to alfalfa in California (Whisson *et al.* 1999), crop damage by African elephants in Kenya (Thouless & Sakwa 1995), Cameroon (Tchamba 1996) and Zimbabwe (Hoare 1999), and fox predation of lambs in Scotland (White *et al.* 2000). Some have forecast outbreaks of rodents, such as *Mastomys* species in Tanzania (Leirs *et al.* 1996) and house mouse in southern Australia (Pech *et al.* 1999). The results of many case studies of rodent dynamics and production damage, especially in Asia and Africa, were described by Singleton *et al.* (1999a), Stenseth *et al.* (2003) and Davis *et al.* (2004b). Comprehensive lists of the extent and value of damage to production by rodents and birds were given by Conover (2002).

Table 5.1: Principles of wildlife damage control for production in agriculture, forestry and fisheries

Economics
5.1: Diminishing returns. Each unit increase in control effort will produce slightly less marginal benefit than the previous unit increase in effort
5.2: Optimal pest control. Optimal pest control occurs when the marginal benefit equals the marginal cost of control
5.3: Discounting. The economics of control of a pest species with a rate of increase (*r*) higher than the discount rate (i), will be different from the economics of controlling a species with a rate of increase lower than the discount rate
Biological community
5.4: Other species. The effect of one species on production depends on the effect of other species

Studies of the effects of pests on livestock may focus on competition for pasture or on predation. Examples of the latter include predation by coyotes in Colorado (Andelt 1992), by golden jackal (*Canis aureus*) in Israel (Yom-Tov *et al.* 1995), by wolf in southern Europe (Meriggi & Lovari 1996), by painted hunting dog (*Lycaon pictus*) in Zimbabwe (Rasmussen 1999), by wolverine in Norway (Landa *et al.* 1999) and by red fox in southern Australia (Greentree *et al.* 2000). Damage by vertebrates to production forests can also occur, such as by red deer in Scotland (Clutton-Brock & Albon 1989) and by grey squirrel (*Sciurus carolinensis*) in Britain (Bryce *et al.* 1997). Predation by birds at aquaculture facilities can be significant, such as in the eastern US (Avery *et al.* 1999; Glahn *et al.* 1999). Production activities may assist the spread of potential pests, such as horticulture possibly spreading exotic frogs in Hawaii (Kraus *et al.* 1999).

These long lists of the range of commodities damaged and the pest species involved around the world are fascinating. This chapter focuses on the principles underlying the great range, especially the economic evaluation of damage control. In theory, this is easier in a production setting than in a conservation or health setting. Examples of economic analyses include the benefit/cost ratio of hunting coyote to reduce sheep predation in parts of Utah and Idaho (Wagner & Conover 1999), the benefit/cost ratio of trapping rats in rice crops in Asia (Singleton *et al.* 1999b) and the economic value of lambs killed by foxes on some farms in Scotland (White *et al.* 2000). Rats cause damage to oil palm and rice crops in Malaysia, but the economic value of the losses is usually not quantified (Wood & Fee 2003). It is surprising that it is easy to find assessments of damage to crops and livestock but hard to find examples with comprehensive economic assessment of the effects of alternative control options.

There are four principles of damage control for production (Table 5.1). These build on the general principles described in Chapters 2 and 3, and the framework illustrated in Figures 1.3 and 1.4 (pp. 8–9). The principles are grouped into those concerned with economics and that concerned with the biological community.

Economics

Production of food and fibre are activities based on economics and ecology. This section discusses fundamental principles of production economics and their application to wildlife damage control.

Diminishing returns

It is well known that increasing inputs such as fertiliser or weed control into a production system, such as crop or livestock production, does not mean that the system can continue to respond in the same way. This is the principle of diminishing returns (the Yield line in Fig. 1.4, p. 9) of production economics (Malcolm *et al.* 1996). In the context of wildlife damage control, the inputs could be trapping, shooting, poisoning, repellents or fertility control. The **diminishing returns** principle states that each unit increase in control effort will produce slightly less marginal benefit than the previous unit increase in effort (Table 5.1).

Note that the yield relationship, such as illustrated in Figure 1.4, encapsulates the net effect of damage and compensatory responses. The niche hypothesis in ecology can be represented as a curve (Krebs 2001) and the diminishing yield curve in Figure 1.4 is a truncated niche curve, showing the left-hand half of the curve. The diminishing returns are the outcome of the **damage response determinants** (Table 2.1, p. 16) and **pest reduction** (Table 3.1, p. 39) principles.

The concept of diminishing returns also occurs in other relationships. For example, as rook abundance in north-east Scotland increased there was an increase in the yield loss of oat and barley crops (Feare 1974) but the loss increased slower than expected (Fig. 2.3a, p. 24).

Optimal pest control

Identifying an optimal level of pest control is a central part of marginal analysis in production economics (Malcolm *et al.* 1996). With low pest abundance and high levels of pest control, it may cost more to input an extra unit of control effort than is returned by the increased production or yield.

The optimal level of shooting of feral pigs in the sheep rangelands of southern Australia (to control predation of lambs) was evaluated by marginal analysis (Choquenot *et al.* 1996). They concluded that shooting, from a helicopter, should continue until the cost per feral pig killed was $90. At that amount the marginal cost equalled the marginal benefit of control (Fig. 5.1). The benefit/cost ratios of various feral pig control strategies were assessed (Choquenot *et al.* 1996) and described by Olsen (1998). The aim was to economically control lamb predation by feral pigs. The strategy that reduced pig abundance to low levels and kept them there had a higher benefit/cost ratio (4.71) than trying to eradicate the pigs (3.96) or maintaining pigs at moderate abundance (3.54).

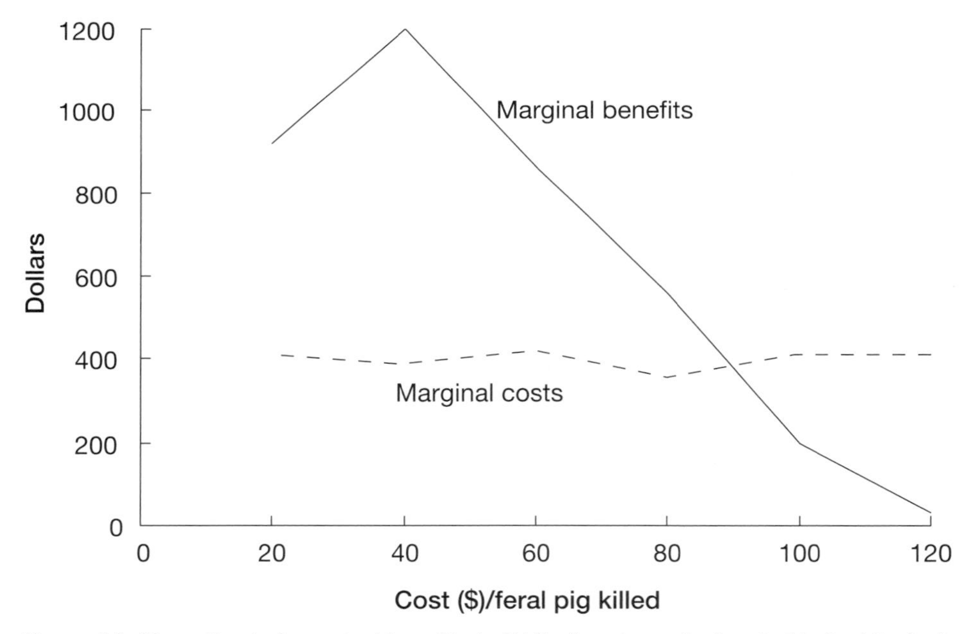

Figure 5.1: The estimated marginal benefits (solid line) and marginal costs (dashed line) of shooting feral pigs to control lamb predation in the sheep rangelands of southern Australia. The estimated optimal level of shooting is when shooting costs $90 per kill, when the marginal benefit equals the marginal cost.

Source: After Choquenot *et al.* (1996).

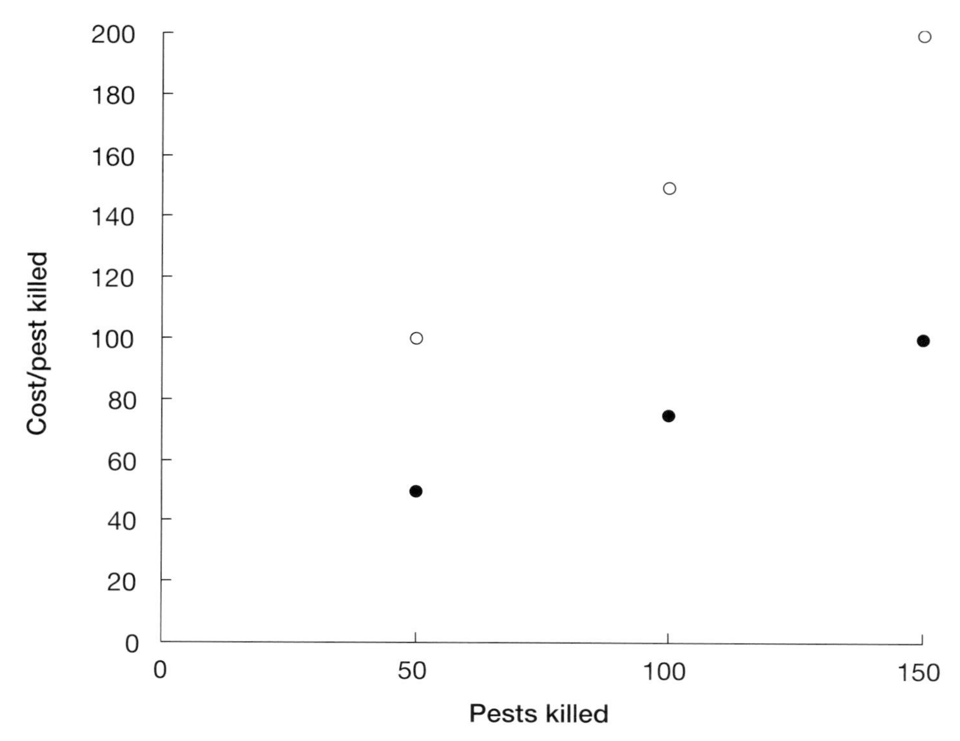

Figure 5.2: The cost-effectiveness of six methods of vertebrate pest control. The most cost-effective methods are shown by solid circles and the least cost-effective by open circles.

The cheapest control for a given objective can be assessed by cost-effectiveness analysis (Fig. 5.2) (Bicknell 1993; Hone 1994a; Cullen *et al.* 2005), which is used primarily when the benefits of control cannot be estimated. Identifying the cheapest method of control does not necessarily mean identifying the optimal control method and its level of use.

The **optimal pest control** principle states that optimal pest control occurs when the marginal benefit equals the marginal cost of control (Table 5.1). This principle is obtained by combining the **diminishing returns** (Table 5.1) and **marginal response** (Table 3.1, p. 39) principles.

The response of yield to damage control may be influenced by the varying amounts of damage caused by different pests (the **individual heterogeneity** principle; Table 2.1, p. 16) and which individuals are controlled (the **susceptibility to control** principle; Table 3.1). For example, the observed variation in crop damage caused by individual Asian elephants and in predation by individual coyotes, lynx and feral pigs may determine the economic responses to damage control. The optimal level of pest control can also be influenced by the relative effectiveness and efficiency of alternative control methods. An evaluation of the alternative methods can include an economic assessment of the effects of substituting one control method for another. This is an example of the principle of **substitution** in production economics (Malcolm *et al.* 1996), noted in Chapter 3 (Table 3.1).

Discounting

The principles of economics are relevant when deciding whether, or how much, to control a pest. This section discusses economics and its relation to pest control.

The future value (A) of an amount of money of current value PV is given by:

$$A = PV(1 + i)^t \tag{5.1}$$

where the interest rate (i) is a compound rate, so the interest is added to the original amount each unit of time (t). Equation 5.1 is the standard equation for calculating the value of money in a bank account when the interest rate is a compound interest rate. The focus is on the future value of a current amount. Discounting focuses on the reverse, the current value of some future amount (Malcolm *et al.* 1996). It involves rearranging equation 5.1 to give:

$$PV = A/(1 + i)^t \tag{5.2}$$

which estimates the present value (PV) of an amount A at time t assuming a discount rate of i, but the terminology changes to correspond to the back-calculation compared with the forward calculation. For example, the future value of \$614 invested with a compound interest rate of 5% ($i = 0.05$) over 10 years is \$1000 (rounded to the nearest dollar). Therefore, solving equation 5.2, the present value of \$1000 (= A) in 10 years time ($t = 10$) assuming a discount rate of 5% ($i = 0.05$) is \$614 (= PV). The discount rate can reflect individual or social perspectives. The social discount rate is often lower than the individual rate (Stenseth *et al.* 2003).

The equation for estimating the future value of a current cost or benefit (equation 5.1) and the equation for exponential population growth are similar. Equation 5.1 is used for estimating future value of a current amount and the equation for exponential population growth is:

$$N_t = N_0 e^{rt}$$

where the future population size is N_t, the current population size is N_0, the instantaneous rate of increase is r and time is t.

Both equations estimate the future value of some current value: A is analogous to N_t, PV to N_0, and t is identical in both equations. This is clearer when e^r is replaced by the finite rate of population growth (λ), so that the equation for exponential growth is then written as:

$$N_t = N_0\lambda^t$$
$$\lambda = e^r$$

Compound interest produces the same pattern of increase in the amount of money as exponential growth does for growth of a population. The similarities are further explored by Berryman (1999).

The relationship between the rate at which a population increases (r) and the rate used to estimate the present value of future costs and revenues (i, the discount rate) has been discussed (Clark 1976, 1981; May 1976; Lande *et al.* 1994) in relation to harvested populations, but it can be generalised to pest species and control of their damage. If the rate of increase is higher than the discount rate ($r > i$), the population can be harvested (controlled) economically. If the rate of increase is less than the discount rate ($r < i$), the best economic strategy is to harvest (control) abundance to very low levels then invest elsewhere the profits earned. Bomford *et al.* (1995) demonstrated that including discounting in the economic evaluation of pest control could change the benefit/cost ratio to less than 1.0, hence control could become uneconomic. An example of discounting in evaluation of control of black-tailed prairie dogs (*Cynomys ludovicianus*) in the US was described by Collins *et al.* (1984). Note the discussion in Collins *et al.* (1984) and in Miller *et al.* (1994) about whether there is clear evidence of prairie dogs' economic effects on livestock production, and Stapp's (1998) discussion of their ecological effects.

These empirical and economic results suggest the **discounting** principle, which states that the economics of control of a pest species with a rate of increase (r) higher than the discount rate (i) will be different from the economics of controlling a species with a rate of increase lower than the discount rate (Table 5.1).

Biological community

More than economics are involved when considering wildlife damage control for production. It also involves species in a community. A final principle examines this point.

Other species

The level of damage by a predator or a herbivore may be partly determined by the availability of alternative food. The original food source may not involve damage; the predator or herbivore may cause damage only after switching food sources. For example, there is a significant negative relationship between the occurrence of domestic animals in the diet of wolves in southern Europe and the occurrence of wild ungulates in the wolf diet (Meriggi & Lovari 1996). This suggests predation of livestock may be highest when availability of wild food is lower. Blackbirds commonly damage corn in parts of North America, though they also eat insects that themselves damage the corn (Dolbeer 1999). The switching between food types can generate a type III functional response in foraging ecology (Begon *et al.* 1996; Krebs 2001).

These empirical and theoretical results suggest the **other species** principle, which states that the effect of one species on production depends on the effect of other species (Table 5.1) in the biological community.

Conclusion

This chapter has described four principles of wildlife damage control for production, and illustrated their links to the framework described in Chapter 1. The next chapter examines principles for managing human and animal health.

Worked examples

1 Marginal analysis

Assume that a crop is being damaged and that the farmer could set traps in and around the crop to control rodents that are eating it. The farmer knows the value of the crop, the value of damage and the yield reduction caused by the rodents and can estimate the cost of trapping. The crop output (in dollars) at different levels of trapping and the costs of trapping are given in Table 5.2. What is the optimal level of trapping? Assume the costs shown in Table 5.2 are variable costs; fixed costs are not shown. The exercise illustrates the **diminishing returns** and the **optimal pest control** principles (Table 5.1).

Solution

The total outputs ($) and costs ($) are shown in Table 5.2 and Figure 5.3. Total outputs of crop 1 increase as trapping effort increases –

Table 5.2: Crop outputs in relation to different numbers of traps used in and around a hypothetical crop. The difference between benefits (total outputs) and costs, and the benefit/cost ratios are also shown. Crop 2 offers a comparative set of total and marginal output data, demonstrating a different pattern of total and marginal outputs

Traps	Cost of traps ($)	Marginal cost ($)	Total output ($)	Marginal output ($)	Benefit-cost ($)	Benefit/ cost ratio
Crop 1						
0	0	–	200	–	200	–
1	200	200	450	250	250	2.25
2	400	200	950	500	550	2.38
3	600	200	1500	550	900	2.50
4	800	200	1950	450	1150	2.44
5	1000	200	2250	300	1250	2.25
6	1200	200	2500	250	1250	2.08
7	1400	200	2700	200	1300	1.93
8	1600	200	2850	150	1250	1.78
9	1800	200	2950	100	1150	1.64
10	2000	200	3000	50	1000	1.50
Crop 2						
0	0	–	400	–	400	–
1	200	200	1000	600	800	5.00
2	400	200	1500	500	1100	3.75
3	600	200	1900	400	1300	3.17
4	800	200	2200	300	1400	2.75
5	1000	200	2400	200	1400	2.40
6	1200	200	2500	100	1300	2.08
7	1400	200	2550	50	1150	1.82
8	1600	200	2590	40	990	1.62
9	1800	200	2620	30	820	1.46
10	2000	200	2640	20	640	1.32

Note: A dash indicates the value could not be estimated.

maximum output ($3000) occurs when 10 traps, the maximum trapping effort, are used. However, increasing trapping from 9 to 10 traps costs $200 to gain extra output worth only $50 (the **diminishing returns** principle). Clearly, the extra cost was greater than the extra benefit. The level of trapping effort at which the extra cost equalled the extra benefit was 7 traps. As trapping effort increased up to 7 traps, the extra benefits exceeded the extra cost. Beyond 7 traps the extra cost exceeded the extra benefit (the **optimal pest control** principle).

The benefits of growing crop 2 were different (Table 5.2, Fig. 5.4). In particular, as the number of traps increased there was an increase in total outputs but a decrease in marginal outputs. Crop 1 showed an increase then a decrease in marginal outputs. Such a difference could be associated

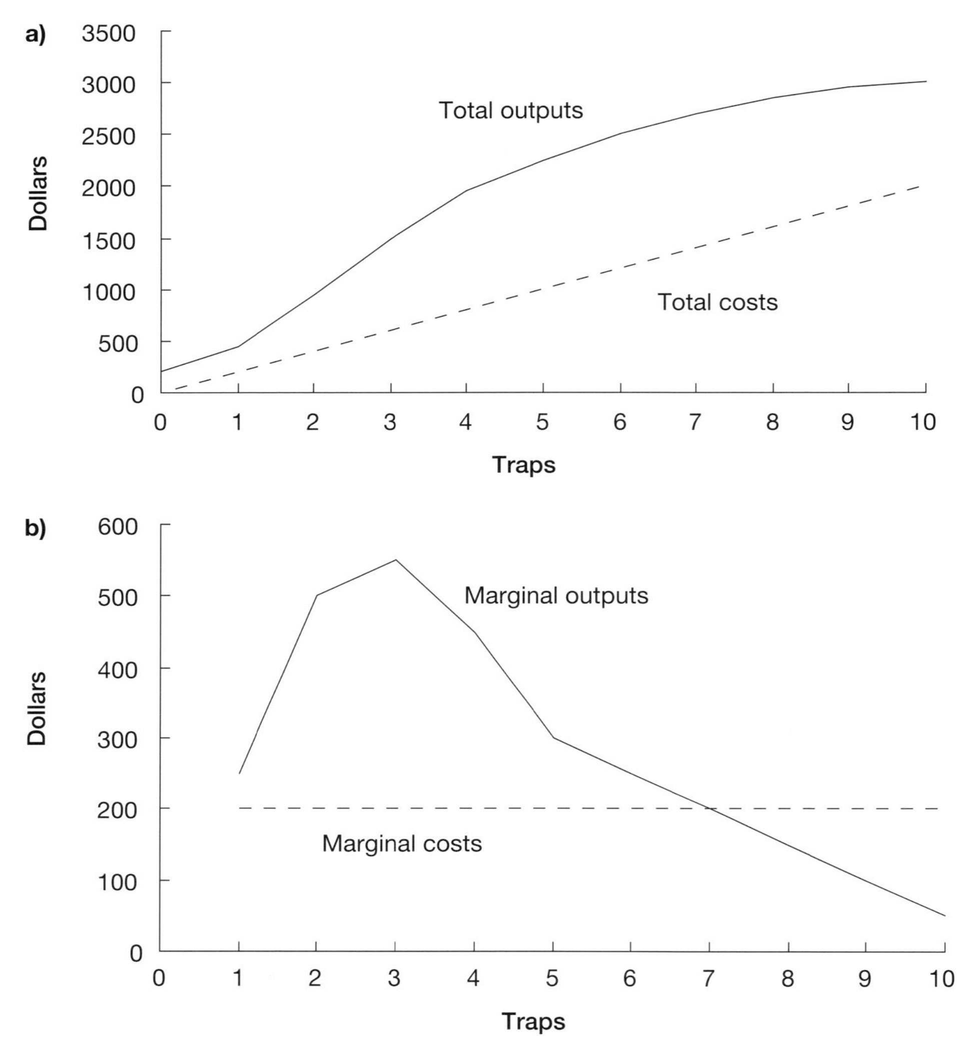

Figure 5.3: (a) The total outputs (solid line) of a crop and the total costs (dashed line) of trapping rodents in and around the crop. (b) The marginal outputs (solid line) of a crop and the marginal costs (dashed line) of trapping. Data are for crop 1 in Table 5.2.

with a compensatory change in the growth as trapping effort increased and presumably rodent abundance and damage decreased. The optimal level of trapping in and around crop 2 was 5 traps.

In each crop the benefit/cost ratio was greater than 1.0, indicating a benefit from control. However, the ratios and analysis of marginal benefits and costs do not produce the same results. In particular, the analysis of benefit/cost ratios does not identify the optimal level of control. The

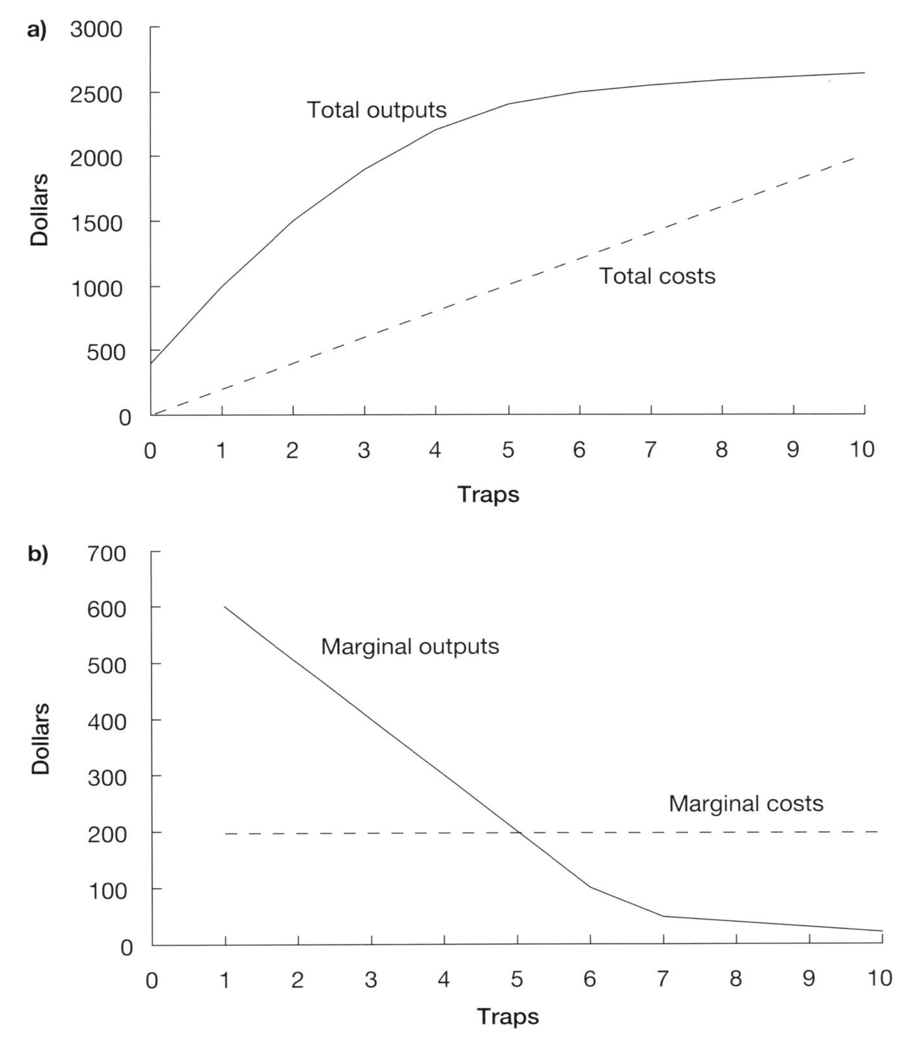

Figure 5.4: (a) The total outputs (solid line) of a crop and the total costs (dashed line) of trapping rodents in and around the crop. (b) The marginal outputs (solid line) of a crop and the marginal costs (dashed line) of trapping. Data are for crop 2 in Table 5.2.

results in this worked example are an extension of some useful economic analyses of trapping of rodents in rice crops in south-east Asia (Singleton *et al.* 1999b).

2 Discounting

Calculate the present value of future costs and benefits of pest control using the data in Table 5.3. The data are for a hypothetical program that

Table 5.3: The benefits and costs of pest control, discounted to present values, using a 5% and a 10% discount rate

	Years		
	1	2	3
Benefits (a)	$10 000	$10 000	$10 000
Costs (b)	$5000	$5000	$5000
Net benefits (c = a – b)	$5000	$5000	$5000
Present value factor (5% discount rate) (d)	0.95	0.91	0.86
Present value factor (10% discount rate) (e)	0.91	0.83	0.75
Present values (f = c*d when 5% rate)	$4750	$4550	$4300
Present values (f = c*e when 10% rate)	$4550	$4150	$3750
Net present value with 5% rate = $4750 + $4550 + $4300 = $13 600			
Net present value with 10% rate = $4550 + $4150 + $3750 = $12 450			

Notes: The calculations estimate the value of dollars at the end of each year, so data for year 1 are for the end of that year. In line d, the present value factor estimated with a 5% discount rate is 0.95 (= $1/(1 + 0.05)^1$) in year 1, and in year 2 is 0.91 (= $1/(1 + 0.05)^2$).

yields gross benefits of $10 000 per year and costs $5000 per year. Assume a 5% discount rate and, for comparison, a 10% discount rate. What is the effect of incorporating a discount rate into the calculations? What is the effect of using a higher discount rate? The exercise illustrates the **discounting** principle (Table 5.1).

Solution

The calculations are shown in Table 5.3. Incorporating a discount rate decreases the current value of future net benefits. A higher discount rate decreases the net benefits even further. A dollar received today is worth more than a dollar received next year, because the dollar received today can be used to earn money between now and next year. For other examples see Malcolm *et al.* (1996).

6

Human and animal health

Introduction

Many wild vertebrates have diseases that also infect humans and/or domestic livestock (Elton 1931; Wobeser 1994; Grenfell & Dobson 1995; Daszak *et al.* 2000; Hudson *et al.* 2001). Examples include foxes, raccoons (*Procyon lotor*) and skunks (*Mephitis mephitis*) infected with rabies, deer and wild pigs infected with foot and mouth disease, wild pigs infected with classical swine fever, badgers, brushtail possums, ferrets and deer infected with bovine tuberculosis (TB), rodents infected with bubonic plague, and elk (*Cervus elaphus*) and bison (*Bison bison*) infected with brucellosis. In Ontario, treatment of humans exposed to rabid wildlife, investigation of exposures and education costs over $3 million annually (Rosatte *et al.* 1997). Control of wild vertebrate diseases in humans or livestock can require vaccination, exclusion fencing or population control of the wild animals.

Diseases may occur in populations of conservation concern (Cleaveland *et al.* 2001). Examples include canine distemper in black-footed ferrets (*Mustela nigripes*) and rabies in painted hunting dogs (also known as African wild dogs). Strategies for control of diseases may involve vaccination of wildlife and vaccination of domestic animals, and limits on movement to reduce contacts (Cleaveland *et al.* 2001).

Table 6.1: Principles of damage control for human and animal health

Host–disease dynamics
6.1: Transmission heterogeneity. There is heterogeneity between hosts in the rate of disease transmission and/or susceptibility
6.2: Spatial spread. The spatial rate of disease spread is related to the population density of susceptible hosts (pests)
Disease control
6.3: Disease reduction. Disease control occurs when the factors that increase the basic reproductive rate (R_0) are reduced
6.4: Vaccination. The proportion of a host (pest) population to vaccinate to reduce disease incidence or prevalence is positively related to the basic reproductive rate of the disease
6.5: Strategic disease control. Strategic disease control involves identifying and limiting the major sources of variation in disease transmission between hosts (pests)
6.6: Multi-host disease control. Controlling a multi-host disease requires controlling the disease in the host populations in which the disease is endemic ($R_0 \geq 1$)
6.7: Disease vectors. Reducing the number of vectors per individual host (pest) will reduce disease incidence or prevalence of a vector-spread disease
Other health issues
6.8: Adverse contacts. Adverse contacts between wildlife and people can be decreased by reducing the abundance of one or both, and changing the behaviour of wildlife and people

Humans and animals often collide with machines such as cars, trucks and aeroplanes. The problem is not new, as illustrated by the discussion in Leopold (1933). The collisions can cause death or injuries and cost millions of dollars annually (Romin & Bissonette 1996; Putman 1997). Bird strikes caused 33 human deaths and cost nearly US$500 million in damage to US Air Force aircraft during 1986–99 (Lovell & Dolbeer 1999). In western Europe, eight species of ungulates were involved in collisions with traffic (cars, trucks and trains) (Groot Bruinderink & Hazebroek 1996). The estimated annual number of collisions with wildlife in Europe was 507 000, involving 300 human deaths, 30 000 injuries and US$1 billion material damage.

Shark attacks on humans cause injury and death. Between 1919 and 1986 there were 27 fatal attacks by sharks in waters off Queensland (Paterson 1990). Attacks also occur elsewhere, such as off South Africa (Wintner & Dudley 2000). In response, efforts are made to catch and exclude sharks from beaches (Gribble *et al.* 1998; Dudley *et al.* 1998).

There are eight principles for control of human and animal health and safety (Table 6.1). They are grouped into three topics: host–disease dynamics, disease control and other health issues. The generic framework illustrated in Figures 1.3 and 1.4 (pp. 8–9) applies here – the response variables are disease incidence or prevalence.

Host–disease dynamics

Host populations are dynamic and so is the interaction of hosts and their diseases. Two principles of host–disease dynamics are examined.

Disease transmission

Many mathematical models of disease spread assume that the rate at which hosts acquire infection is proportional to the product of the population densities of susceptible hosts and infectious hosts (Anderson & May 1991; McCallum *et al.* 2001; Begon *et al.* 2002). This assumption involves 'density-dependent transmission'. Another assumption is 'frequency-dependent transmission', which assumes that the rate of disease transmission is proportional to the density of susceptible hosts and the ratio of infectious hosts to total density (McCallum *et al.* 2001). The observed rates of disease transmission are estimated by various models of the type described by McCallum *et al.* (2001) and Fenton *et al.* (2002).

Some hosts may transmit more pathogens, or be more likely to get infected by a disease, because of differences in behaviour, age or genotype (Swinton *et al.* 2001; Begon *et al.* 2002). Such heterogeneity can be incorporated into models of disease dynamics, such as those for bovine TB in badgers in England (Smith *et al.* 1997, 2001) and brushtail possums in New Zealand (Barlow 2000). A very common pattern with macroparasites is that some hosts have no or few worms and a few hosts have high worm burdens (Wilson *et al.* 2001). There is heterogeneity between hosts in the rate of disease transmission and/or susceptibility. This is the **transmission heterogeneity** principle (Table 6.1).

Spatial spread

Mathematical models of spatial disease spread by diffusion predict that the rate of spread is proportional to the population density of susceptible hosts. An example is the spatial model of rabies in foxes (Kallen *et al.* 1985). Different models make different assumptions about a positive relationship (Kallen *et al.* 1985), or positive then negative relationship (van den Bosch *et al.* 1990). Disease may also be spread between 'patches' of hosts in a manner analogous to movement of individuals between parts of a metapopulation (Grenfell & Harwood 1997). This can predict local extinction but global persistence of disease. The theoretical results suggest the **spatial spread** principle, which states that the spatial rate of disease spread is related to the population density of susceptible hosts (pests) (Table 6.1).

Disease control

The control of undesirable diseases is assisted by knowing the effects of human actions on hosts and their diseases. Five principles of disease control are examined.

Disease reduction

Control of an undesirable disease requires management, which may involve reducing host density by vaccination or culling. If host density drops or is pushed below the threshold density (K_T), the disease cannot establish or persist in the host population. For example, removing skunks (by lethal control) may have reduced rabies prevalence in Alberta (Rosatte *et al.* 1986). Reducing the density of brushtail possums in New Zealand (Caley *et al.* 1999) was associated with a decline in gross prevalence of bovine TB in the possums (Fig. 6.1). In these two studies the reduction was of total density, not solely density of susceptible hosts. However, the conclusions of both studies were limited by the lack of experimental controls, as noted by the authors. Vaccination of raccoons against canine distemper in Ontario led to a significant reduction in the prevalence of the disease (Schubert *et al.* 1998b) compared with an experimental control area (Fig. 6.2).

If there is a relationship between prevalence and density, reducing the population density of susceptible hosts should reduce disease prevalence. Evidence of such a relationship is consistent with the expectation, but the evidence is not compelling. That requires an actual reduction in susceptible host density and observation of the expected response. A positive relationship between prevalence of brucellosis and abundance of bison in Yellowstone National Park was described by Dobson and Meagher (1996). There was an estimated threshold abundance (K_T or N_T) of about 200 bison, above which prevalence was greater than zero (Fig. 6.3). Those results were derived from a model of brucellosis–bison dynamics assuming density-dependent transmission of the pathogen. If transmission was

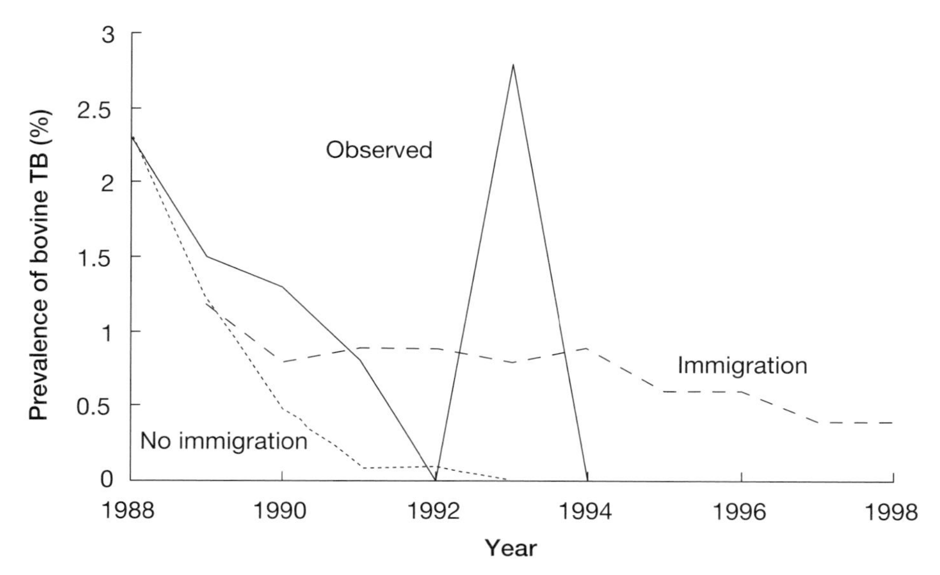

Figure 6.1: Observed trends (solid line) in prevalence (%) of bovine TB in brushtail possums after possum control. Also shown are predicted trends when immigration of possums occurs (dashed line) and when no immigration occurs (dotted line).

Source: After Caley *et al.* (1999).

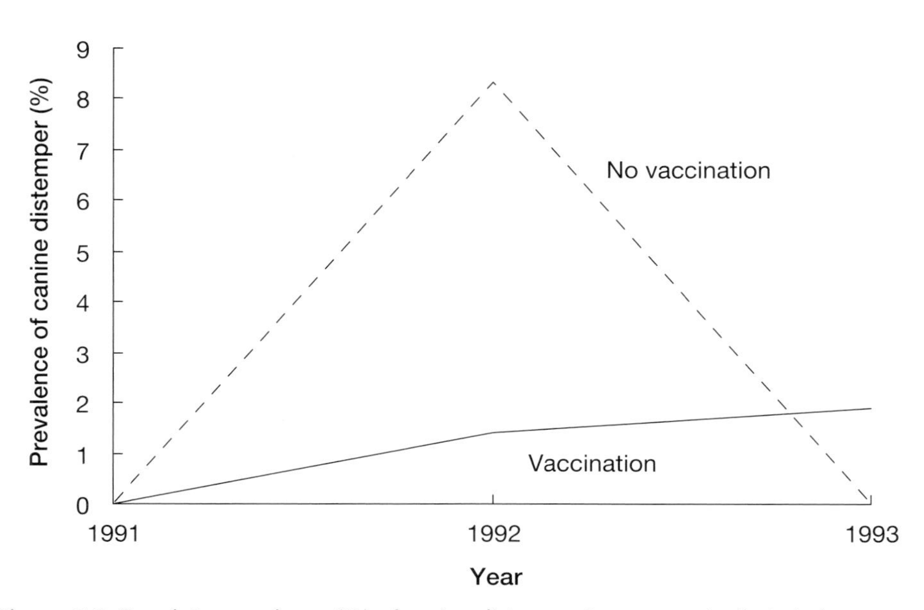

Figure 6.2: Trends in prevalence (%) of canine distemper in raccoons in Ontario in an area with vaccination (solid line) and no vaccination (dashed line).

Source: After Schubert *et al.* (1998b).

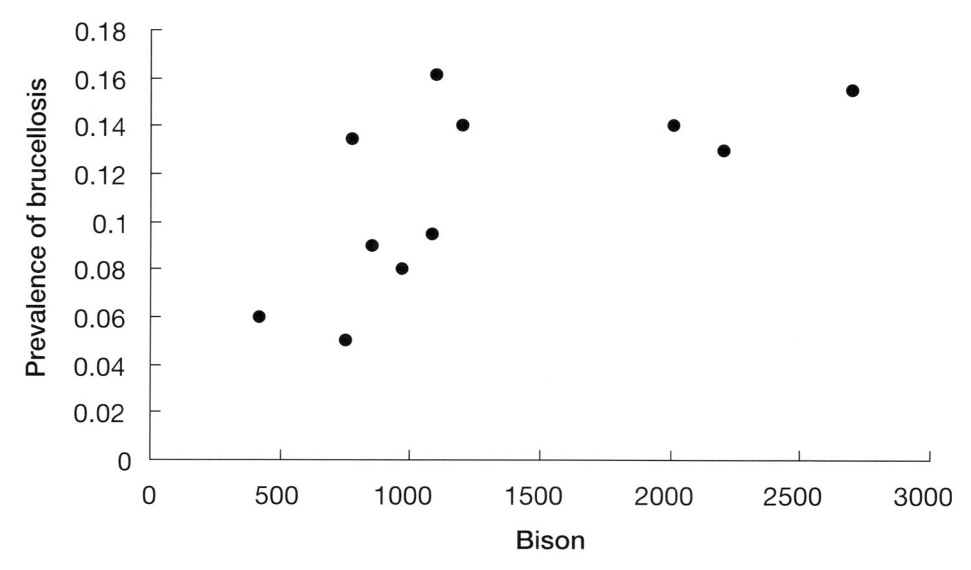

Figure 6.3: The relationship between prevalence as a proportion of brucellosis in bison, and bison abundance in Yellowstone National Park.

Source: After Dobson and Meagher (1996).

frequency-dependent then there was no relationship between prevalence and abundance, and hence no threshold.

The threshold is analogous to the minimum viable population (MVP) in the **threshold population parameters** principle (Table 4.1, p. 69) (see Grenfell & Harwood 1997). The threshold density of rats for establishment of bubonic plague was estimated as 3000 km^{-2} (Keeling & Gilligan 2000). The likelihood of detecting plague in Kazakhstan increased above a threshold abundance of great gerbil (*Rhombomys opimus*) (Davis *et al.* 2004a). The threshold density of wild boar for establishment of classical swine fever was estimated in a deterministic model as 6.1 km^{-2} (Hone *et al.* 1992). The threshold abundance of feral pigs for establishment of porcine transmissible gastroenteritis was between 161 and 805, depending on the assumed transmission coefficient (Hone 1994b). The threshold abundance of feral pigs to foot and mouth disease was estimated to be higher in a stochastic than in a deterministic model, for semi-arid Australia (Dexter 2003). The source of stochasticity was variable food availability because of variable rainfall.

The basic reproductive rate (R_0) of an infection is the average number of secondary infections per infected individual in a population of susceptible hosts (Anderson & May 1991; Swinton *et al.* 2001). The parameter is analogous to the net reproductive rate (R) in population dynamics. The (illegal) successful introduction of rabbit haemorrhagic disease into New Zealand (Parkes *et al.* 2002) implies $R_0 \geq 1$, in contrast to the unsuccessful (legal and illegal) introductions of myxomatosis into New Zealand (Parkes *et al.* 2002), implying $R_0 < 1$. Disease control occurs when the factors that increase the basic reproductive rate (R_0) are reduced. This is the **disease reduction** principle (Table 6.1). Typically such factors (parameters in disease models) are the transmission rate and, if transmission is density-dependent, susceptible host density.

A complication could arise in control of disease involving a second host species and a vector. Cases of bubonic plague in humans may increase after control of rats (a wildlife host of bubonic plague) as the vectors (fleas) move from dead rats to humans (Keeling & Gilligan 2000). These issues are explored later, in the discussion of the **multi-host disease control** principle.

Vaccination

As the basic reproductive rate (R_0) of a disease increases, above a threshold of 1.0, the proportion (p) of hosts we need to vaccinate to stop disease spread also increases (Fig. 6.4), but at a slower rate (Anderson & May 1982, 1991), as described in equation 6.1. For a discussion of the estimation of R_0 see Dietz (1993).

$$p = 1 - (1/R_0) \qquad (6.1)$$

The proportion of foxes in Europe to vaccinate to control (reduce the incidence of) rabies was estimated by:

$$p = 1 - (K_T/K) \qquad (6.2)$$

where K_T is the threshold fox density (no./km^{-2}) and K is carrying capacity (no./km^{-2}) (Anderson *et al.* 1981). As K increases then p increases (Fig. 6.5). In Ontario it was estimated that 46–80% of foxes were vaccinated (Rosatte *et al.* 1992), which would be too low if K was high and K_T was about 1 fox/km^{-2}. Equation 6.1 assumes that vaccination occurs at random in the host population. If there are differences between hosts in the rate at which they spread disease (the **transmission heterogeneity** principle, Table 6.1), equation 6.1 may give biased estimates. For example, if infection was concentrated in 20% of the host population then vaccination of that 20% would have an effect different

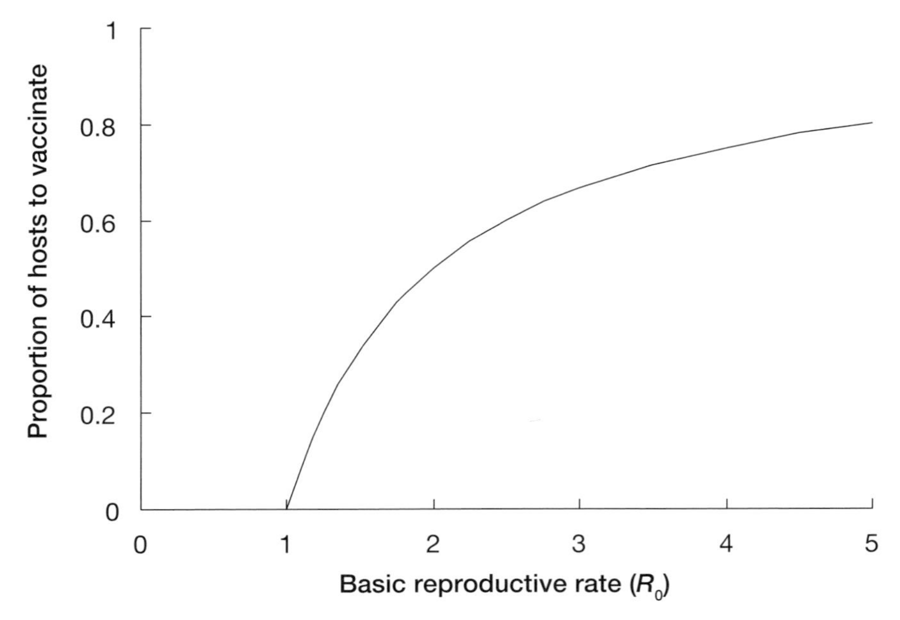

Figure 6.4: The relationship between the basic reproductive rate (R_0) of an infectious disease and the proportion of susceptible hosts to vaccinate to stop disease spread.

from that predicted by equation 6.1. Conversely, if the 80% of the host population without infection were vaccinated the effect would be different again. Such biases are examined briefly by Randolph *et al.* (2001) in relation to tick-borne infections.

These empirical and theoretical results suggest the **vaccination** principle, which states that the proportion of a host (pest) population to vaccinate to reduce disease incidence or prevalence is positively related to the basic reproductive rate of the disease (Table 6.1).

If the hosts are of conservation concern, care must be taken to ensure the vaccine is safe. There can be disastrous effects if a vaccine is not checked sufficiently. For example, a canine distemper vaccine used to vaccinate black-footed ferrets proved lethal to the ferrets (Thorne & Williams 1988).

Strategic disease control

Classical epidemiological theory assumes random mixing (= density-dependent transmission) of susceptible and infectious hosts (Anderson & May 1979, 1991). Such mixing may not occur in wild pest populations because of social structures, territorial behaviour and non-random movement patterns. For example, bovine TB in brushtail possums in

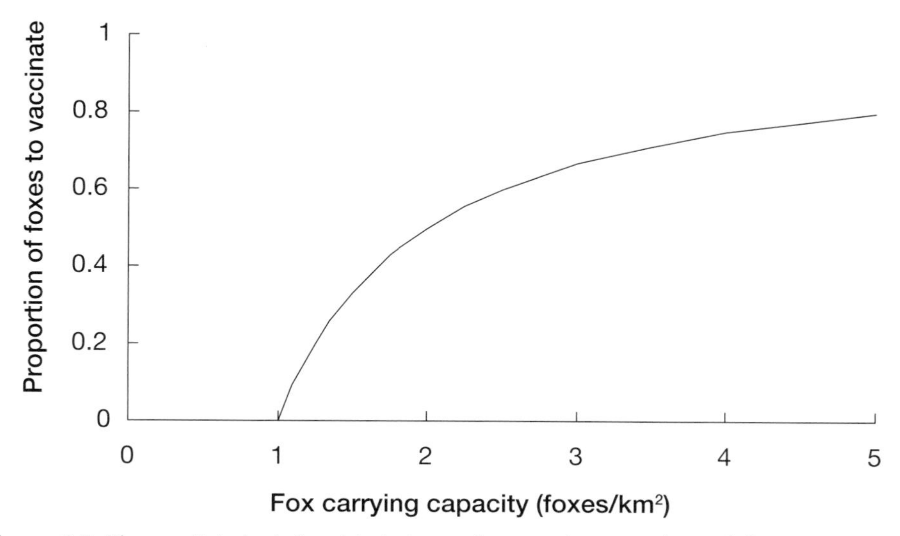

Figure 6.5: The predicted relationship between fox carrying capacity and the proportion of foxes to vaccinate to control rabies.

Source: Modified from Anderson *et al.* (1981).

New Zealand appears to spread in a non-random manner (Barlow 1991). Some infectious individuals in the host population will have a higher probability of contacting susceptible individuals and hence of transmitting the disease (the **transmission heterogeneity** principle, Table 6.1). Strategic control involves identifying those infectious individuals and limiting their chance of disease transmission by changing their behaviour or movements, removing them or vaccinating them. This targeting lowers the proportion of hosts we need to control (cull or vaccinate) to stop disease spread. Such theoretical effects have been described by Anderson and May (1991).

For effective disease control by vaccination, hosts should be vaccinated before they get infected. Hence the average age at vaccination should be less than the average age at first infection (Anderson & May 1982; Williams *et al.* 1996).

The theoretical results suggest a **strategic disease control** principle, which involves identifying and limiting the major sources of variation in disease transmission between hosts (pests) (Table 6.1).

Multi-host disease control

Diseases such as rabies, TB and foot and mouth disease can have many wild host species. For example, bovine TB has been reported in 14 species of wild mammals in New Zealand (Coleman & Cooke 2001) and continued cattle TB infection may require transmission of infection from wildlife (Kean *et al.* 1999). The prevalence of macroscopic bovine TB infection in ferrets was positively correlated with brushtail possum abundance and was unrelated (Caley 1998) or weakly related (Caley *et al.* 2001) to ferret abundance in New Zealand. Control of the disease in host species 1 may have little or no effect on disease incidence in host species 2. This could occur if host species 1 is a dead-end host ($R_0 = 0$) or a spill-over host ($0 < R_0 < 1$). Control of the disease will be necessary in host species where the disease is endemic ($R_0 \geq 1$), because in such hosts the disease can establish and persist.

Cases of rabies in wildlife in Tanzania's Serengeti National Park may be related to the disease in domestic dogs in nearby areas, when dog density exceeds 5 km^{-2} (Cleaveland & Dye 1995). It was predicted that control of wildlife rabies may be achieved in such areas by vaccination of more than 70% of domestic dogs. Oral vaccination of foxes against rabies has occurred in Ontario (Schubert *et al.* 1998a). Those authors considered that such vaccination may reduce the incidence of rabies in foxes, which may then reduce the incidence in skunks because of reduced transmission between species. Control of badgers in parts of Ireland was associated with a reduction in bovine TB in cattle in the area (O'Mairtin *et al.* 1998) (Fig. 6.6). Sustained reduction (by lethal control) of populations of brushtail possums was associated with significantly reduced incidence of bovine TB in domestic cattle in part of New Zealand (Caley *et al.* 1999). Population control of brushtail possums resulted in a significantly lower prevalence of bovine TB in ferrets (Fig. 6.7) in New Zealand (Caley 2001; Caley & Hone 2004). Control of badgers in south-western England reduced prevalence of bovine TB in the badger population but did not stop infection in cattle in the area (Tuyttens *et al.* 2000). Localised badger control may have had little effect on bovine TB incidence in cattle in south-western England and may even have increased it (Donnelly *et al.* 2003). A subsequent study reported badger culling reduced cattle TB incidence in the culled area but increased incidence in adjoining areas (Donnelly *et al.* 2006).

The empirical and theoretical results suggest a **multi-host disease control** principle, which states that controlling a multi-host disease requires controlling the disease in the host populations in which the disease is endemic ($R_0 \geq 1$) (Table 6.1).

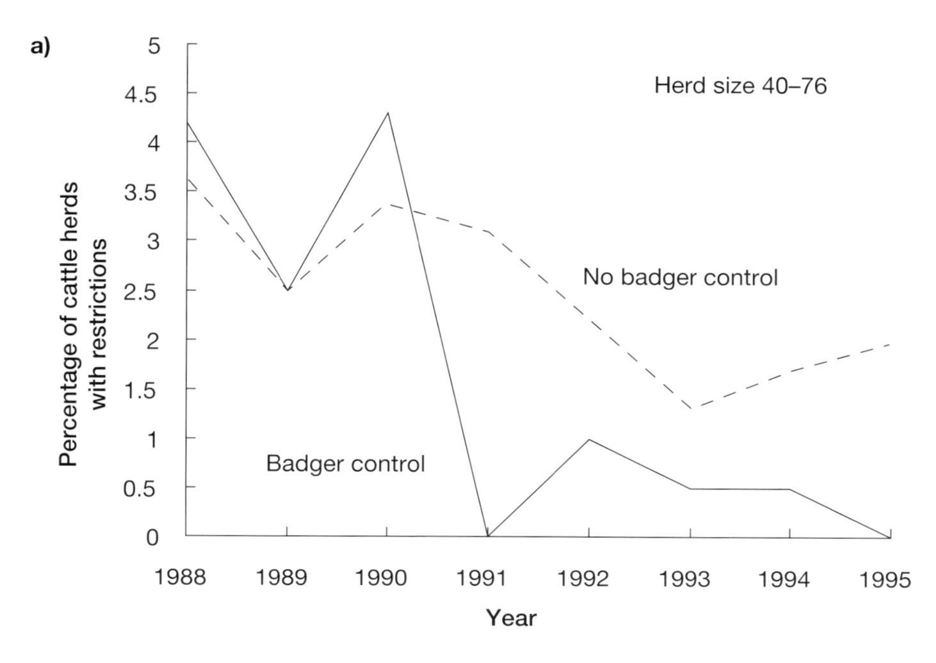

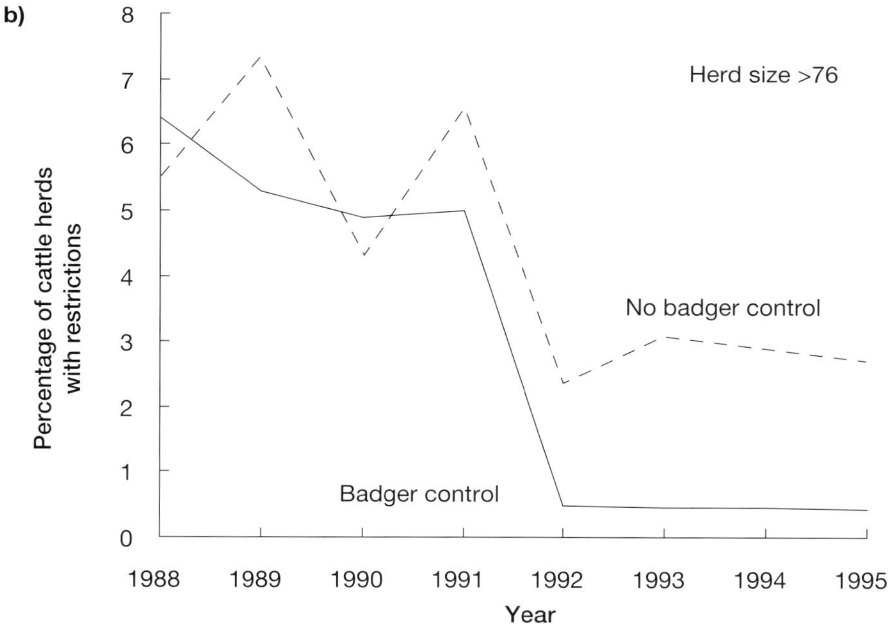

Figure 6.6: Trends in the percentage of cattle herds of different sizes that had restrictions (detection of at least one TB reactor during the year) in an area with badger control (solid line) and no badger control (dashed line) in Ireland. (a) Herds of 40–76 cattle. (b) Herds of 76+ cattle.

Source: After O'Mairtin *et al.* (1998).

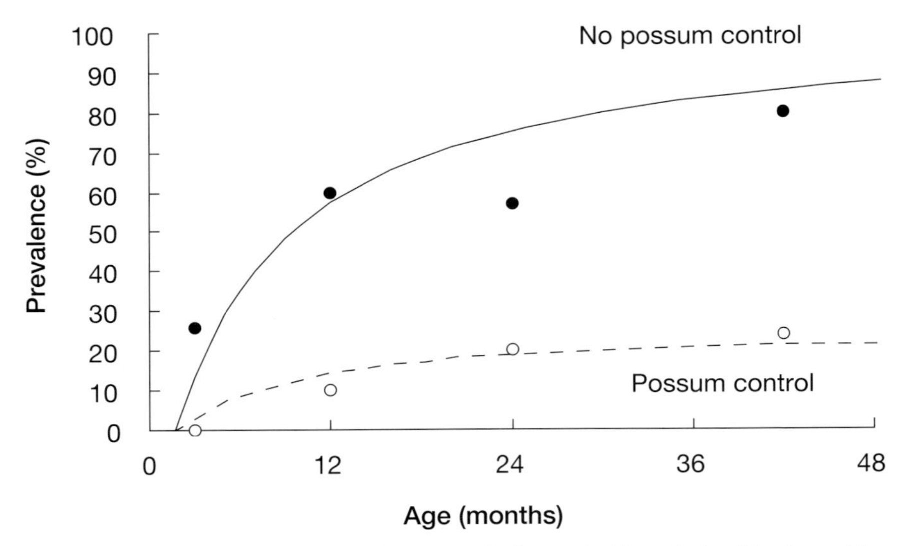

Figure 6.7: Age-specific prevalence of bovine TB in ferrets in New Zealand in sites with no possum control (solid circles and solid line) and sites with possum control (open circles and dashed line). The lines are best-fit regressions with parameters estimated across sites.

Source: Modified from Caley (2001).

Disease vectors

Some diseases, such as plague and Japanese encephalitis, are spread from host to host by a species of insect. This is called the disease vector. Epidemiological theory predicts the number of secondary infections per infected host (the basic reproductive rate) is proportional to the ratio of vectors per host (May & Anderson 1979; Anderson 1984; Gupta *et al.* 1994). One study in England found that the abundance of fleas, a vector of myxoma virus, was reduced by use of an insecticide, and rabbit abundance increased (Fig. 6.8) (Trout *et al.* 1992). Reducing tick abundance is a strategy to control Lyme disease in deer and humans (Randolph *et al.* 2001). The introduction of myxomatosis into New Zealand may have been unsuccessful because there was a lack of appropriate vectors (Parkes *et al.* 2002).

The empirical and theoretical results suggest the **disease vectors** principle, which states that reducing the number of vectors per individual host (pest) will reduce disease incidence or prevalence of a vector-spread disease (Table 6.1).

Other health issues

The collisions between wildlife, humans and machines can be usefully explored using ecological patterns and processes. This section examines

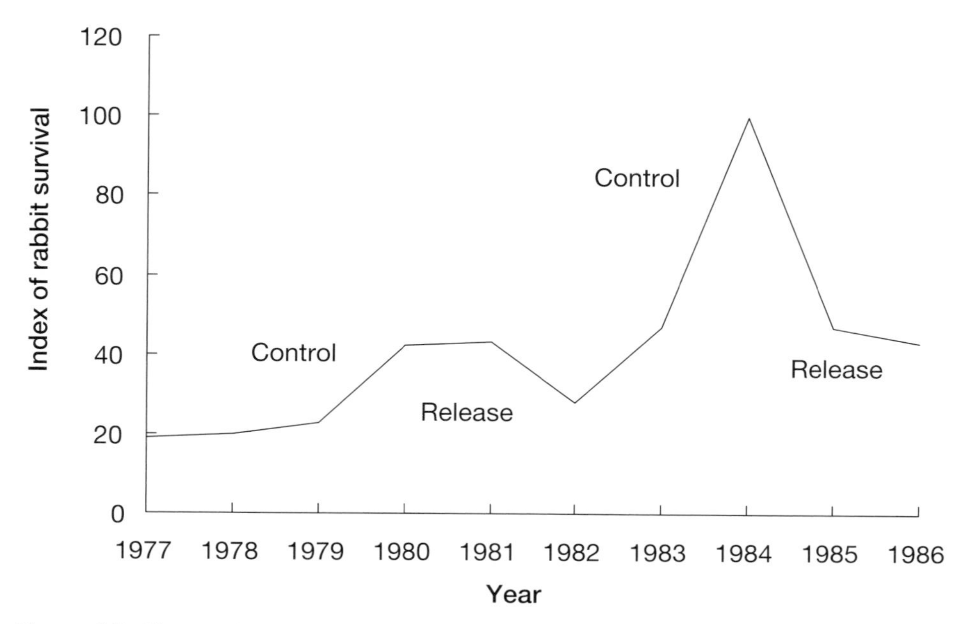

Figure 6.8: Changes in over-winter survival of rabbits in Hampshire, during years of control (1978–80, 1982–84) of abundance of rabbit flea, a vector of myxomatosis, and years of release (1980–82, 1984–86) of rabbit fleas.

Source: After Trout *et al.* (1992).

these from the perspective of a relationship in predator/prey theory and harvesting theory in ecology.

Adverse contacts

Collisions between wildlife and machines, such as cars, trucks and aeroplanes, are analogous to harvesting or predators catching prey accidentally. Reductions in the number of collisions can theoretically be achieved by reducing animal abundance (N), reducing machines (E) or changing the behaviour of animals or drivers. Shark attacks on people are a different form of wildlife–people coincidence. Harvesting theory and predator/prey theory assume that the number of animals caught (n) is proportional to the product of the number of animals available to be caught (N) and effort (E) (Begon *et al.* 1996). This was expressed as equation 3.1, repeated here:

$$n = aNE \tag{6.3}$$

where a is the proportionality constant. Many other forms of the relationship can occur (Hone 1994a). Groot Bruinderink and Hazebroek (1996) cited earlier work of a positive correlation between abundance of five species of ungulate (N) and road kills (n). A significant positive relationship, through the origin, occurred between weekly road kills of kangaroos (n) in western New South Wales and weekly night-time traffic (E) (Klocker *et al.* 2006).

Collisions between kangaroos (*Macropus giganteus*) and cars/trucks occurred at similar rates on roads with and without warning signs in southern Australia (Coulson 1982). Such signs are an attempt to reduce the coefficient a in equation 6.3. Efforts to reduce collisions between deer and cars/trucks focus on increasing driver awareness (reduce a in equation 6.3), reducing the number of deer road-crossings (reduce N), changing the pattern of crossings (reduce N) (Putman 1997), changing animal behaviour (reduce a) (Ujvari *et al.* 1998) and identifying the reasons for the heterogeneity in locations of collisions (reduce a) (Hubbard *et al.* 2000). In Europe, a combination of fencing and wildlife passage was recommended to reduce collisions on roads and railways with high traffic volumes and high speed limits (Groot Bruinderink & Hazebroek 1996).

Efforts to reduce shark attacks on people involve exclusion netting (reduce a) and baited traps (reduce N) (Dudley *et al.* 1998; Gribble *et al.* 1998). However, these efforts may injure or kill other marine species such as whales, dugong, turtles and dolphins (Paterson 1990; Gribble *et al.* 1998). Such unintended effects are examples of the **response to control** principle (Table 3.1) – the occurrence of non-target effects of pest control.

The empirical and theoretical results suggest the **adverse contacts** principle, which states that adverse contacts between wildlife and people can be decreased by reducing the abundance of one or both, and changing the behaviour of wildlife and people (Table 6.1).

Conclusion

This chapter has examined principles in the management of human and animal health. The principles focus on identifying and using knowledge of the determinants of disease and host dynamics, and factors influencing the coincidence of wildlife, humans and machines. The next chapter describes wildlife damage control and human recreation.

Worked example

1 Parameter values and their application

Describe the similarities and differences between the parameters R_0, R, λ and the benefit/cost ratio. When using any of the parameters for management of pest damage, describe the level of change in each parameter needed to achieve a particular management objective. The example illustrates the **disease reduction** principle (Table 6.1) and the **threshold population parameters** principle (Table 4.1, p. 69).

Solution

The basic reproductive rate of a disease (R_0) is the average number of secondary infections caused by an infected host in a population of susceptible hosts (Anderson & May 1991). Net reproductive rate (R) is the average number of female young per female per generation in a population (Krebs 2001). Hence R_0 is a property of a disease and R is a property of a host population. Lambda (λ) is the finite rate of population growth (Caughley & Sinclair 1994). The benefit/cost ratio is the estimated benefits of a management action divided by the estimated costs of the management.

The three parameters R_0, R and λ have the similar property that a numerical value of 1.0 corresponds to a stable state, a value of 0 corresponds to extinction, and a value greater than 1.0 corresponds to growth in disease incidence (R_0) or host population size (R, λ) respectively. The benefit/cost ratio has a key value of 1.0 when benefits equal costs, suggesting there is no net benefit. When the ratio is less than 1.0 then the action costs more than it returns, and when the ratio is greater than 1.0 then management has a positive net benefit. Hence, the numerical value of 1.0 for each parameter can be a guide to the outcome or consequences of management actions.

The proportional change (p) in each parameter needed to change a value from greater than 1.0 to equal to 1.0 is given by:

$$p = 1 - (1/x)$$

where x is any of the four parameters. This is simply a generalised version of equation 6.1 and the relationship shown in Figure 6.4. Hence, if $x = R_0$ then p equals the proportional reduction in the average number of secondary infections per infected host. If management achieves a reduction greater than p, disease prevalence or incidence decreases (disease control occurs). If $x = R$ then p equals the proportional reduction in average number of female young per female per generation. If management achieves a reduction greater than p, host (pest) density decreases (pest control occurs).

7
Recreation

Introduction

People have a variety of attitudes to wildlife. Some people see wildlife as sources of food or fur, others see wildlife as animals to appreciate through binoculars or a camera lens. To some people, wildlife is for hunting. Other people disregard wildlife or fear it. The removal of animals from the wild, for food or as a hunting trophy, is called 'consumptive recreation'. The use of wildlife without such removal, such as wildlife viewing, is termed 'non-consumptive recreation'. The effects of non-consumptive recreation can be positive or negative for wildlife (Boyle & Samson 1985; Green & Higginbottom 2000).

Visiting coastal areas to observe seabirds, such as puffins, penguins and nesting terns, is an example of non-consumptive recreation. In some locations, for example coastal Victoria and the South Island of New Zealand, people visit penguin colonies. Introduced mammalian predators are controlled in and around the colonies to serve both human recreation opportunities and penguin conservation.

Wildlife, such as deer and geese, can damage golf courses, parks and gardens by what they eat; wildlife, such as pigeons and geese, can foul statues and ponds with their excreta (Feare 1991; Dolbeer *et al.* 1994; VanDruff *et al.* 1994). A survey of people in large metropolitan areas in the US reported that 61% suffered a wildlife-related problem (Conover

1997a). The problems may represent US$250 million in damage to households (Conover 1997b).

If the animals in parks and gardens are having undesirable effects, action may be needed to reduce or stop those effects. Mass population control is usually not an option in urban areas, particularly with large vertebrates such as deer, geese or kangaroos. Population management is more likely to involve removal of individuals considered or known to cause the most problems, or use of repellents to reduce damage without reducing abundance. An alternative method is fertility control, investigated for white-tailed deer in urban areas (Nielsen *et al.* 1997; Rudolph *et al.* 2000; Rutberg *et al.* 2004). Another form of population management involved the use of wires to exclude rock doves (*Columba livia*), and their undesirable fouling, from ledges at a football stadium in Colorado (Andelt & Burnham 1993). An example of a repellent is the use of bird feeders in the UK that cannot be accessed by grey squirrels, which otherwise eat the seeds provided for birds.

The need for population management also occurs outside urban areas and towns. For example, hikers and campers in national parks and wilderness areas in Canada and the US may encounter bears – either a thrill or a dangerous pest, depending on the frequency of encounters (Kerr & Wilman 1988). Encounters with snakes may cause a similar range of reactions. At picnic grounds in parts of the western US, signs warn that rodents in the area carry plague (Conover 2002). The Australian magpie can be viewed as a bird with a musical call, or as a dangerous pest for attacking people. Peregrine falcons can be viewed as a magnificent raptor or as a pest that kills domestic racing pigeons. Holiday-makers with young children on Fraser Island, Queensland, are warned of the abundance of dingoes (*Canis lupus dingo*) and what to do if a dingo approaches too closely. These warnings follow a fatal attack on a child in 2001.

In some recreational areas used for hunting (consumptive recreation), predators may be thought to kill prey which the hunters consider their game, and the hunters call for the predators to be killed. For example, in Britain, buzzards (*Buteo buteo*) kill some ring-necked pheasants (*Phasianus colchinus*) at or near sites where pheasants are released for shooting (Kenward *et al.* 2001). Some predators are controlled to increase the harvest of game. For example, red fox and hen harrier (*Circus cyaneus*) are controlled in areas of Scotland managed for hunting of red grouse (*Lagopus lagopus*) (Green & Etheridge 1999).

Table 7.1: Principles relating to wildlife damage and the management of recreation

Damage
7.1: Recreation damage. There are relationships between pest abundance, habitat features and damage to recreation
7.2: Damage heterogeneity. Individuals in a wildlife population cause differing amounts of damage to recreation
7.3: Damage variation. Damage to recreation varies spatially and temporally
Damage control
7.4: Control response. The level of damage reduction in response to control is determined by pest abundance and control efforts
7.5: Control heterogeneity. Control affects individuals differently
7.6: Unintended effects of control. There may be unintended effects of control on recreation

Predator control was viewed as the second of five steps in a developing sequence of traditional game management (Leopold 1933). The steps were:

- restriction of hunting;
- predator control;
- reservation of game lands;
- artificial replenishment (now often called releases or stocking);
- environmental controls.

Predator control was often based on assumed (not demonstrated) predator–game relationships. Such assumptions were not encouraged (Leopold 1933).

These wide-ranging examples may not involve a conservation issue but there are issues of wildlife detracting from human recreational activity, including our aesthetic appreciation of wildlife and other components of a landscape. The principles for management of wildlife damage relative to recreation are grouped into two topics – damage, and damage control (Table 7.1). These are mostly applications of principles from earlier chapters, which are described where appropriate. There is considerable overlap between the content of this chapter and earlier chapters, reflecting the similarities in wildlife damage issues. However, here the emphasis is on the effects of wildlife on recreation. The principles described in this chapter may be hypotheses rather than principles demonstrated repeatedly in empirical studies. This reflects the limited field research into human recreation and wildlife damage control.

Damage

The following three principles of wildlife damage control focus on the relationship between wildlife and its effects on human recreation, and factors that contribute to variation around that relationship.

Determinants of damage

Hikers may walk to high-elevation swamps and grasslands in parts of south-eastern Australia and find, to their surprise and dismay, that feral pigs have converted the native vegetation and landscape to one resembling a ploughed paddock. Examples of such effects on vegetation were described by Hone (2002). Wild horses occur in mountainous areas of south-eastern Australia (Dyring 1990; Walter 2002); they create wide tracks and erode streambanks. The tracks cause a decrease in plant cover and an increase in bare soil (Dyring 1990). Such effects are similar to those produced by people, described by Liddle (1997). The horses also leave large piles of dung on tracks designed for recreational walkers. In each of these situations, the recreational experience has been diminished.

A common, usually implicit, assumption is that more animals cause more problems (more damage). For example, with more pigeons there is more excreta on statues and ledges of buildings. With more geese there is more damage to golf courses. Mathematically, the assumption can be expressed as a positive relationship, maybe linear, maybe curved. VanDruff *et al.* (1994) stated that generally there was a poor relationship between damage and density of urban wildlife, though no data or analyses were presented to support that view.

The relationship between visitor numbers and wildlife abundance, or demographic rates, has been little studied. In one study of cliff-nesting birds in a reserve in Scotland, the breeding success of kittiwakes (*Rissa tridactyla*) and guillemots (*Uria aalge*) was negatively related to the number of visitors and the average distance between birds and people (Beale & Monahan 2004). It was suggested that people were analogous to non-killing predators. Where predators kill prey the effect is likely to be related to predator abundance and the per capita functional response of each predator. These issues were discussed in Chapter 2. Predators may also affect prey through harassment and causing the prey to forage in areas with lower food quantity and quality (Leopold 1933; Hik 1995; Arthur *et al.* 2004). Obviously, visitors to a seabird colony are not actual predators but the analogy is useful to help plan management of real predators and their effects on human recreation.

In the absence of many empirical studies estimating relationships, the possible relationships between recreation and wildlife can be examined and presented as hypotheses for evaluation. For example, as seabird abundance increases the visitor numbers to an area also increase (Fig. 7.1a). The relationship may be stated as 'at locations with more seabirds there are more visitors'. If the abundance of seabird predators increases and predation also increases, then seabird abundance decreases (Fig. 7.1b). This is an example of the framework relationship illustrated in Figure 1.3a (p. 8). Substituting the relationship in Figure 7.1b into the relationship in Figure 7.1a, we see that as predators increase visitor numbers decrease (Fig. 7.1c). Alternatively, at locations with more predators there are fewer visitors (Fig. 7.1c).

The relationships shown in Figure 7.1 are hypothetical, but can be evaluated. The linear relationship is only one of a range of specific examples of the relationships, as they could vary from simple curves to compound (sigmoidal, parabolic) curves. Also, different habitats, such as sandy coastlines or rocky coastlines, may correspond to relationships where the lines have different slopes or intercepts. The relationship between seabirds and visitors may change if people have negative effects on the seabirds, such as reducing the breeding success of kittiwakes and guillemots. For example, the relationship in Figure 7.1 may become steeper (for the same number of visitors there are fewer seabirds) and curve back towards the y axis (large numbers of visitors associated with very few seabirds).

The limited empirical experience and theoretical ideas suggest the **recreation damage** principle, which states that there are relationships between pest abundance, habitat features and damage to recreation (Table 7.1). This is an application of the framework illustrated in Figure 1.3 (p. 8) and the **damage extent** principle (Table 2.1, p. 16).

Heterogeneity

While more pests can cause more damage, not all pests are equal. Some individuals may contribute more to damage than others. For example, larger (and likely older) saltwater crocodiles (*Crocodylus porosus*) are perceived as a greater problem than young crocodiles in northern Australia. Some Australian magpies attack people at certain times of year, for example while the people are walking through a park. The magpies that attack are nearly all males (Jones 2002), and particular magpies may attack only pedestrians or only cyclists (Warne & Jones 2003). In southern

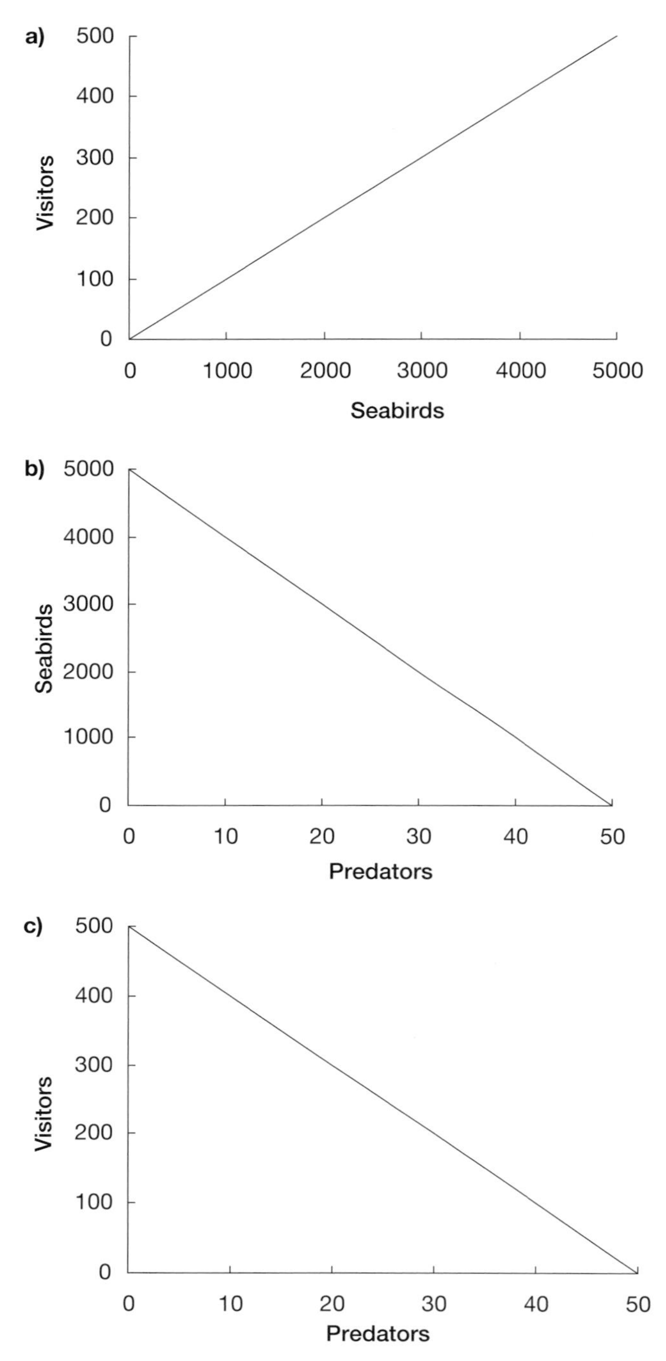

Figure 7.1: (a) A possible relationship between visitor numbers to a coastal location and seabird abundance. (b) A possible relationship between seabird abundance at a coastal location and predator abundance. (c) The derived possible relationship between visitor numbers to a coastal location and predator abundance.

England, buzzards that fledged close to pheasant pens were more likely to visit the pens than were buzzards that fledged further away (Kenward *et al.* 2001). A hypothesis is that the former buzzards may be more likely to attack the pheasants, which are game birds.

Individuals in a wildlife population cause differing amounts of damage to recreation. This is the **damage heterogeneity** principle (Table 7.1), an application of the **individual heterogeneity** principle of Chapter 2.

Spatial and temporal variation in damage

Wildlife affect human recreation at some times of year and in some locations. Attacks by Australian magpies occur during the breeding season and mostly when the magpies have young in the nest (Jones 2002). Predation of game pheasants by buzzards varies spatially, being highest at release pens with little nearby shrub cover (Kenward *et al.* 2001).

Damage to recreation varies spatially and temporally. This is the **damage variation** principle (Table 7.1), an application of the **damage variation** principle of Chapter 2.

Damage control

In planning and executing wildlife damage control, there are many choices. They include education programs to increase awareness of the issue and help people make decisions that will reduce their conflicts with wildlife. For example, management of problem Australian magpies can involve educating people about the risks and responses (Jones & Thomas 1999). These dimensions of human–wildlife conflicts are examined in more detail by Conover (2002) and Jones (2002).

The management of problem wildlife in recreation areas involves selecting one or more methods of damage control from a long list of possible methods. These may include the broad strategic options of lethal or non-lethal control, or finer tactical options such as trapping, repellents, scaring devices, shooting and so on. Each method will involve benefits and costs, and a formal assessment, such as cost–benefit (Hone 1994a), could be useful.

Control response

The fouling of buildings by pigeons and other birds is often controlled by bird removal or exclusion. In a study of urban pigeons that involved removing some birds, there was a change in apparent dispersal and immigration (Sol & Senar 1995). Pigeons redistributed themselves between groups.

Many historical examples of predator control for the benefit of human recreation were described by Leopold (1933). Many studies of predation in ecology have focused on the effects of predator control on game bird abundance. Many such studies in the UK, Europe and North America were reviewed by Newton (1998). In areas of Britain managed for hunting of red grouse, predators of the grouse, such as red fox and hen harriers, are controlled. An alternative approach is to offer the hen harriers supplementary food, to reduce their predation on grouse. Such supplementary feeding reduced predation on grouse chicks in south-western Scotland (Redpath *et al.* 2001). Control of predators, such as red fox, carrion crow (*Corvus corone*) and magpie (*Pica pica*) was associated with increases in breeding success and abundance of grey partridge (*Perdix perdix*) in southern England (Tapper *et al.* 1996). However, prey, such as birds, may not always show clear responses to predator control (Cote & Sutherland 1997). Also, apparent responses may not be actual responses, as discussed for mule deer (*Odocoileus hemionus*) and predator control on the Kaibab North Plateau in the US (Caughley 1970).

When controlling problem wildlife for the benefit of human recreation, a common aim is to reduce damage but not eradicate the wildlife. This is similar to the features of a sustainable harvest strategy, where the aim is to harvest annually but not harvest the population to extinction. Simulation has shown that a feature of sustained harvests is 'escapement' (Lande *et al.* 1994, 1997; Kokko 2001; Milner-Gulland *et al.* 2001). That is, there is a threshold number of animals (abundance or density) below which we do not harvest. This is known as 'threshold harvesting' – the population is not harvested to low levels or to extinction. The strategy can increase the probability of conserving the resource and the harvest. In the context of managing problem wildlife for human recreation, such an escapement strategy would mean not removing all the problem animals.

Returning to the hypothetical coastal seabird colony example illustrated in Figure 7.1, we will examine the effect of predator control on visitor numbers. If higher levels of predator control effort cause a decrease in predator abundance (Fig. 7.2a), it can be hypothesised, by combining relationships in Figure 7.1 and Figure 7.2a, that higher levels of predator

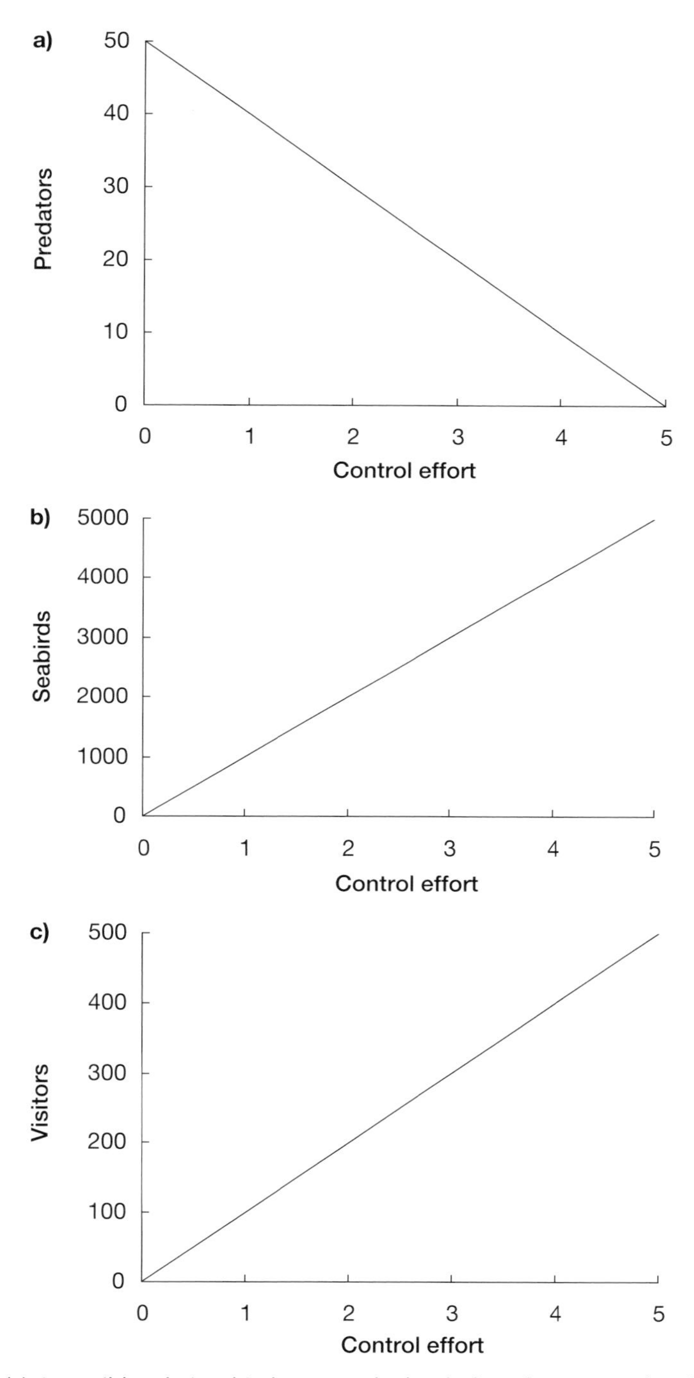

Figure 7.2: (a) A possible relationship between the level of predator control and predator abundance. (b) A possible relationship between the level of predator control and seabird abundance. (c) A possible relationship between the level of predator control and visitor numbers.

control increase seabird abundance (Fig. 7.2b) and visitor numbers (Fig. 7.2c). The relationships are simplified, like all such models, but they can be evaluated empirically in field studies. In the field the relationships are likely to be curves. For example, with higher levels of predator control there are likely to be diminishing returns (the **diminishing returns** principle in Table 5.1, p. 88) so that as effort increases, the marginal response of seabird abundance and visitor numbers is likely to get smaller. Hence, the response by seabirds may be an example of the framework relationship shown in Figure 1.4a (dotted line, p. 9).

The level of damage reduction in response to control is determined by pest abundance and control efforts. This is the **control response** principle (Table 7.1), an application of the **response to control** principle of Chapter 3.

Control heterogeneity

When damage control occurs some individuals can be more affected than others. For example, some saltwater crocodiles near beaches where people swim may be harder to catch than others. Those captured can be relocated, but some return to the site of capture (Walsh & Whitehead 1993). Some Australian magpies may be trapped and relocated (Jones & Thomas 1999) but a few individuals return to their original territory while others do not (Jones 2002).

The empirical experience suggests the **control heterogeneity** principle, which states that control affects individuals differently (Table 7.1). This is an application of the **susceptibility to control** principle of Chapter 3.

Unintended effects of control

When predators are controlled to increase game abundance there may be unintended effects, for example some other prey species may increase in abundance. This is known as mesopredator release (Courchamp *et al.* 1999c). When problem animals such as Australian magpies are trapped and removed from urban parks, there may be unintended effects. Relocated individuals may compete with other magpies already in the area, or with other species, and reduce their abundance.

The empirical experience suggests the **unintended effects of control** principle, which states that there may be unintended effects of control on recreation (Table 7.1). This is an application of the **response to control** principle of Chapter 3.

Conclusion

This chapter has described principles for managing wildlife damage to human recreation. The principles were often applications of those described earlier in other situations, either generally or in discussions of biodiversity conservation or production. The principles are proposed here mainly as hypotheses for future evaluation. Such evaluation is encouraged.

Worked example

1 Harvesting feral goats

A conservation reserve has a population of feral goats, which have been shown to be adversely affecting native vegetation and the recreational experience and enjoyment of visitors. As a result, the managers of the reserve have developed a management plan. The aim is to reduce the effects of feral goats by reducing goat abundance from 500 to 50 goats, and to keep abundance at 50 on an annual basis. The managers consider it is impossible to eradicate the goats because of the high cost of removing the last few goats, as illustrated in Figure 3.12b (p. 61). The managers want to estimate how many goats have to be removed each year, for example by recreational hunters, to keep abundance at 50 goats. To estimate the number of goats to remove annually, it is necessary to describe the dynamics of the goat population. The example illustrates the **control response** principle (Table 7.1).

Examine the effects of assuming the dynamics of the feral goat population can be described by logistic growth, or the numerical response. Assume the intrinsic rate of increase, r_m, is 0.41 per year, based on the results of Maas (1997, 1998) in western New South Wales. The situation is analogous to that of feral goats in Egmont National Park in New Zealand (Forsyth *et al.* 2003), where managers tried to reduce impacts but could not remove all the goats, and of feral horses in coastal areas in the eastern US (Turner 1988; Goodloe *et al.* 1991), where managers did not want to remove all the horses.

Solution

Management of the goats is in two parts – the initial reduction, then the maintenance phase in which goat abundance is kept low. The number (n) of goats to remove each year in the maintenance phase from the population of 50 is given by rN. Assuming logistic growth, the number to remove annually (n) is given by:

$$n = rN = r_m N(1 - (N/K))$$
$$n = 0.41 \times 50(1 - (50/500)) = 18$$

where it is assumed that carrying capacity (K) is 500 goats. Note that 50 goats is well below the abundance ($K/2 = 250$) at which the maximum annual sustained harvest ($= r_m K/4 = 0.41 \times 500/4 = 51$ goats) could be taken.

The number (n) of goats to remove each year, assuming population growth as described by the numerical response of Maas (1997, 1998), is also rN, but now is given by:

$$n = rN = (-a + c(1 - e^{-dV}))N$$

Parameter values of a, c and d were estimated by Maas (1997). When food supply is high ($V = 500$ kg/ha) and the parameter values ($a = 0.85$, $c = 1.264$, $d = 0.0059$) have been inserted, the number to remove is:

$$n = (-0.85 + 1.264 \times (1 - e^{-0.0059 \times 500})) \times 50 = 17$$

which is very similar to that estimated (18) assuming logistic growth.

If food supply (V) is very low, such as when $V = 100$ kg/ha, the annual number to remove is:

$$n = (-0.85 + 1.264 \times (1 - e^{-0.0059 \times 100})) \times 50 = -14$$

which is interpreted as 0. In other words, no goats have to be removed when food supply (V) is very low, such as in a drought, as goat abundance will decline anyway.

The comparative results are shown in Figure 7.3. The estimates of goats to remove each year are similar if food supply (V) is very high but quite different when food supply is low. In particular, when food supply is low removal of 18 goats per year would be an overestimate and would presumably be unattainable, given that managers cannot eradicate the goats. Below a food supply of 189 kg/ha no goat control is needed, as abundance declines because there is not enough food (Maas 1997, 1998). The example illustrates the difference between an equation that assumes a constant environment (logistic equation) and an equation that assumes a variable environment (numerical response equation). An analogous study of red kangaroos was described by Caughley and Gunn (1996).

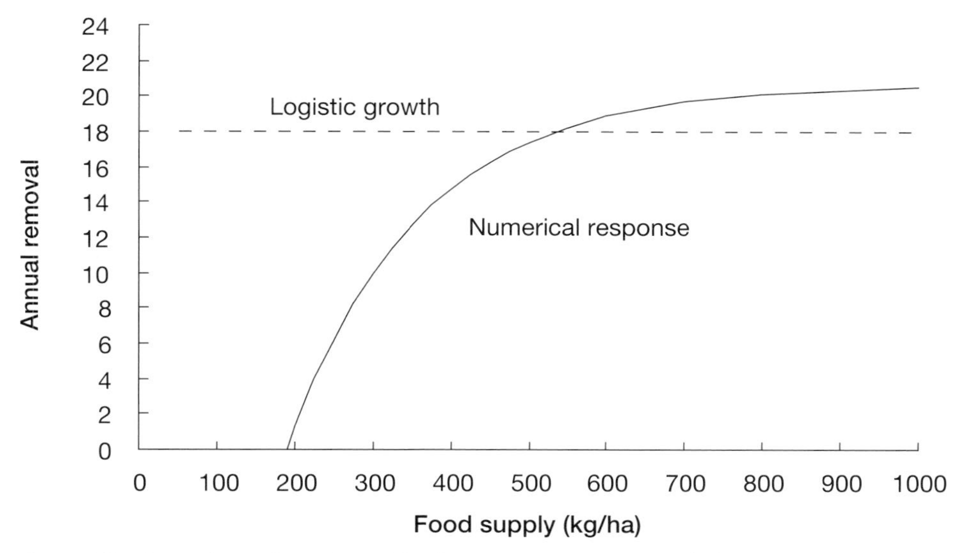

Figure 7.3: The estimated number of feral goats to remove annually from a population to keep abundance at 50 goats. It is assumed that the population dynamics are described by a numerical response (solid line) or by the logistic equation (dashed line).

Feral goat control using hunters in Egmont National Park in New Zealand was evaluated by modelling goat population dynamics using a logistic equation (Forsyth *et al.* 2003), as there were no data on food availability.

The numerical response of *r* on food availability is asymmetric (Caughley 1987; Owen-Smith 2002). When food availability drops below the equilibrium level, population size falls faster than it rises when food availability rises by the same amount above the equilibrium amount. The consequences of such variation were examined by Davis *et al.* (2003), who reported it had important consequences for population dynamics – the variation reduced *r*. In this worked example that implies that fewer feral goats would need to be removed annually than predicted by simple logistic growth (which assumes a constant environment).

To use this approach in the field, it is necessary to be able to estimate food availability (*V*) at a particular time. It is also necessary to know that the population growth rate (*r*) during the subsequent period, such as three months, is related to food availability, as demonstrated by Maas (1997). The results can be linked to other species for which the numerical response on food has been demonstrated, such as red foxes (Pech & Hood 1998), feral pigs (Choquenot 1998), brushtail possums (Bayliss &

Choquenot 2003) and the per capita numerical response of wild horses (Walter 2002). Such numerical responses to rainfall have been demonstrated for feral buffalo (Skeat 1990) and feral pigs (Caley 1993). Other examples are described in Sibly *et al.* (2003).

If adaptive management (Walker 1998) of feral goats occurred, the recreational enjoyment of visitors would be assessed before and after goat control. The results would help estimate whether management was achieving its aims, rather than assessing simply whether feral goat control occurred.

8
Conclusions

The science of wildlife damage control involves many implicit and some explicit generalisations. Those generalisations are presented here as principles and are discussed so that they are clear for all readers to understand and evaluate. The principles are integrated across a range of topics to illustrate key framework relationships (Figs 1.3 and 1.4, pp. 8–9) relevant to different management objectives. The aim of identifying the generalisations is to overcome apparent problems caused by the multitude of possible damage issues, pest species, locations, control methods and times.

The principles listed and discussed are mostly generalisations previously established empirically. Others are based on relevant theory and are proposed here as hypotheses. Evaluation of these is encouraged. They, like all the principles, should not be interpreted as dogma or facts to be accepted without assessment. If generalisations, or principles, 'leak' it is important we learn why, so the science of wildlife damage control can advance.

Readers may agree with some principles and disagree with others. We hope that readers will be stimulated to contribute to the development of a more rigorous and useful science of wildlife damage control. The development of such a science will not be easy and it will take time. There will be many vigorous debates and progress will at times seem frustratingly slow – these are difficult scientific, economic and statistical problems.

A challenge for the future is to get scientists and managers to use the principles as illustrated in Figures 1.3 and 1.4, evaluate them and improve upon them where necessary. Another challenge is to consider multiple objectives of wildlife damage control. For example, a particular species may cause damage to biodiversity and production and health, yet be harvested – the issue is how to assess which objective has priority and the effects of different population densities (Fig. 1.3). Similarly, the many potential effects of increasing management effort (Fig. 1.4) can be evaluated. The outcome of such evaluations will be improved wildlife damage control.

Appendix 1
Introductory mathematics

Several mathematical functions or relationships are described, to assist readers with limited knowledge of mathematics. The key functions are logarithms and exponentials.

Logarithms

Logarithms are a mathematical function between two variables, x and y. Logarithms are used in many activities. For example, if we measure the pH of soil or water, the measure of pH is on a logarithmic scale. The magnitude of an earthquake is measured on the Richter scale, which is a logarithmic scale. An earthquake of magnitude 7 is 10 times bigger than an earthquake of Richter magnitude 6, not 1 unit larger.

In simple terms, logarithms are a positive curved relationship between x and y (Fig. A1.1). There are two common types of logarithmic relationships – logarithms to base 10 (usually called common logarithms, and written as $\log_{10}x$) and logarithms to base e (usually called natural or Naperian logarithms, and written as $\log_e x$ or $\ln x$). These are illustrated in Figure A1.1. When $x = 10$, $\log_{10}10 = 1$. When $x = e$, $\log_e e = 1$. Note that the logarithmic relationships are curved, and pass through the point (1, 0). When x is greater than 1.0, $\log x$ is greater than 0 (is positive) and as x increases $\log x$ increases but at a slower rate. When x is less than 1.0

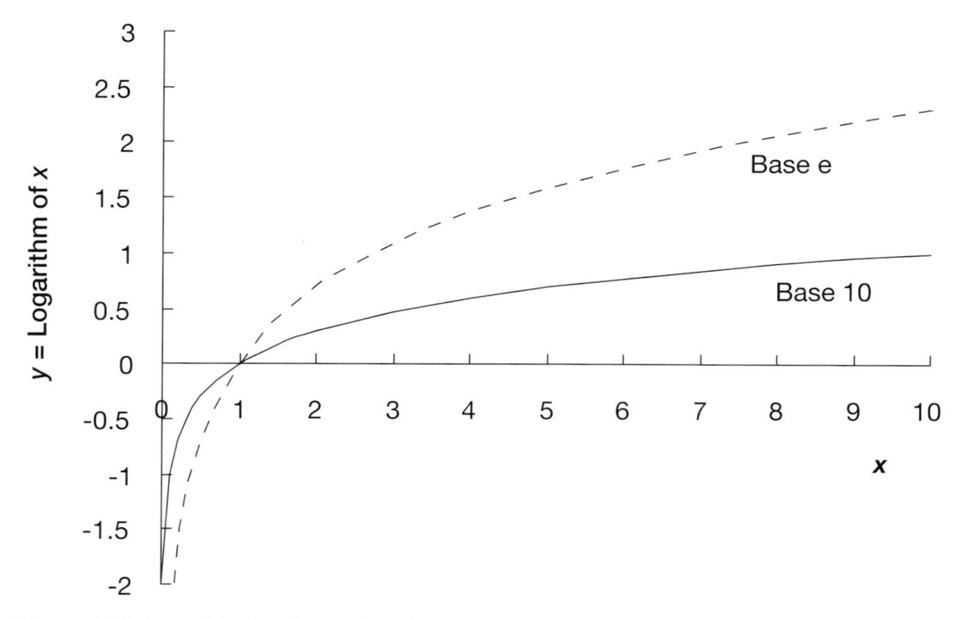

Figure A1.1: Logarithmic relationships between *y* and *x* for logarithms to base 10 (solid line) and to base e (dashed line), for *x* up to 10.

but greater than 0, log x is negative, and as x approaches 0 then log x becomes more negative and at a faster rate. The lines on the graphs for log x approach the y (vertical) axis but never intercept it (Fig. A1.1).

A simple example can illustrate the ideas. The number 10 is equal to 10^1. The number 100 is equal to 10^2, 1000 is equal to 10^3 and so on. On a logarithmic scale to base 10, these numbers correspond to 1, 2, 3 and so on. The logarithm of the numbers 10, 100, 1000 equals the number which is the exponent (the number raised up to the right of each 10).

Logarithmic functions are used in several ways. They are used to transform data, for example in analysis of variance, in order to achieve normality or additivity. Logarithms are also used to convert finite rates to instantaneous rates. These purposes are described below.

The logarithmic function is illustrated below for common applications. Note that the calculations do not require us to specify whether the logarithms are to base 10 or base e, hence the base is not usually shown. However, it is shown when specific calculations occur.

If $y = ax$:

$$\log y = \log a + \log x$$

For example, $8 = 4 \times 2$. Hence:

$$\log 8 = \log 4 + \log 2$$
$$\log_{10} 8 = 0.903 = \log_{10} 4 + \log_{10} 2 = 0.602 + 0.301 = 0.903.$$

If $y = x/a$:

$$\log y = \log x - \log a$$

for example $4 = 8/2$. Hence:

$$\log 4 = \log 8 - \log 2$$
$$\log_{10} 4 = 0.602 = \log_{10} 8 - \log_{10} 2 = 0.903 - 0.301 = 0.602.$$

If $y = x^a$:

$$\log y = \log x^{\mathrm{a}} = a \log x$$

for example, $8 = 2^3$. Hence:

$$\log 8 = \log 2^3 = 3 \log 2$$
$$\log_{10} 8 = 0.903 = \log_{10} 2^3 = 3 \log_{10} 2 = 3 \times 0.301 = 0.903.$$

If $y = ax^b$:

$$\log y = \log a + \log x^b = \log a + b \log x$$

for example, $8 = 1 \times 2^3$. Hence:

$$\log 8 = \log 1 + \log 2^3 = \log 1 + 3 \log 2$$
$$\log_{10} 8 = 0.903 = \log_{10} 1 + 3 \log_{10} 2 = 0 + 3 \times 0.301 = 0.903.$$

A common application of logarithms is to transform data prior to analysis such as for linear regression. The equation $y = ax^b$ is commonly called a power function. The equation is illustrated in Figure A1.2 for various values of the exponent b. One method of estimating the value of the parameters a and b in the power function is to transform the equation into a linear regression by using logarithms. For example, after transformation the equation is:

$$\log y = \log a + \log x^b = \log a + b \log x$$

Then the regression is done by assuming $Y = \log y$ and $X = \log x$. The estimated slope equals b and the estimated intercept equals $\log a$. Hence, to obtain an estimate of the slope on an arithmetic scale not a logarithmic scale, the estimate is back-transformed. That is, if the intercept

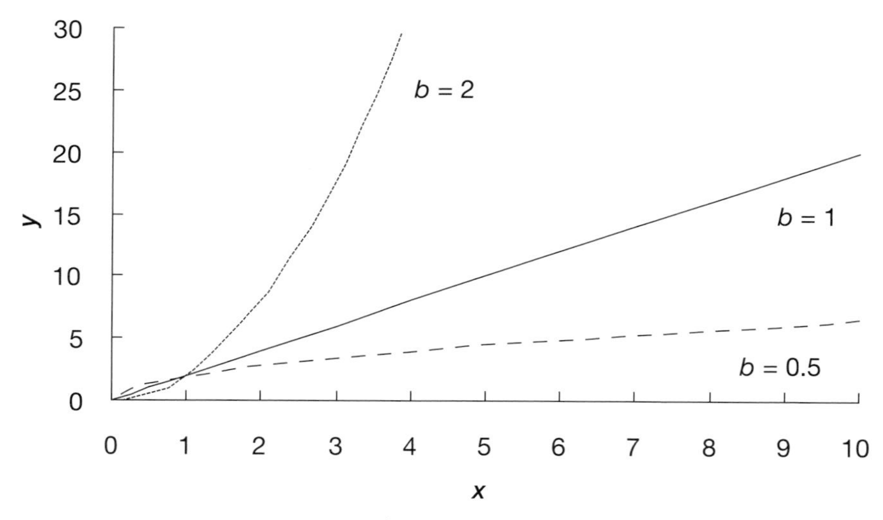

Figure A1.2: The power function, $y = ax^b$, for different values of the exponent b: $b = 0.5$ (dashed line), $b = 1$ (solid line) and $b = 2$ (dotted line). In each equation $a = 2$.

equals $\log a$ then, assuming the logarithms are to base 10, $10^{\text{intercept}} = a$.

Exponentials

Exponentials are related to logarithms. For example, $100 = 10 \times 10 = 10^2$. Also, $\log_{10}100 = \log_{10} 10^2 = 2\log_{10} 10 = 2 \times 1 = 2$. Hence the logarithm, to base 10, of 100 equals the value of the exponent (2). In population ecology, an important exponential function uses the number e. The number e is defined as:

$$e = 1 + (1/1!) + (1/2!) + (1/3!) + \ldots + (1/n!) = 2.71828\ldots$$

where 1! is called one factorial ($= 1 \times 1$), 2! is called two factorial ($= 2 \times 1$), 3! ($= 3 \times 2 \times 1$) and so on.

The number e is important in modelling population dynamics. Assuming the instantaneous rate of population growth ($\mathrm{d}N/\mathrm{d}t$) is proportional to population size (N):

$$\mathrm{d}N/\mathrm{d}t \;\alpha\; N$$

hence $\mathrm{d}N/\mathrm{d}t = rN$ where r is the proportionality constant, and is called the instantaneous population growth rate or rate of increase. Integration of the equation gives:

$$N_t = N_0\,e^{rt}$$

where population size at time t is N_t and at time 0 is N_0. The equation describes exponential population growth over time. The equation can also be written as:

$$N_t = N_0\,\lambda^t$$

where λ is the finite population growth rate or rate of increase. The parameters r and λ are linked as $r = \ln \lambda$ and $\lambda = e^r$.

Additive and multiplicative effects

If two variables determine the value of a third variable, there are several possibilities as to how the first two variables are related mathematically. For example, the number of pests killed (n) by shooting may be related to the number of pests (N) and the shooting effort (E). That is:

$$n = f(N, E)$$

where $f(N, E)$ describes an unspecified mathematical function of N and E. If N and E are related in an additive manner then:

$$n = aN + bE$$

which is an additive multiple linear regression with coefficients a and b, which are to be estimated. It is implicit that the intercept in the equation is 0. Alternatively, if N and E are related in a multiplicative manner then:

$$n = aNE$$

where the coefficient a is to be estimated. That is, the value of N is multiplied by the value of E to estimate n.

With such alternative equations it is important to evaluate their mathematical strengths and weaknesses by checking their behaviour at the extremes of possible data values. For example, what is the value of n when $N = 0$? The additive equation solves to $n = bE$ when $N = 0$. This does not make sense, as no animals can be killed by shooting when there are no animals to shoot. The multiplicative equation solves to $n = 0$ when $N = 0$, which does make sense. The same occurs for situations with no control effort ($E = 0$). The additive equation is rejected in favour of the multiplicative model.

The concepts of additive and multiplicative relationships are linked to the use of logarithms. For example, if $n = aNE$, after taking logarithms of both sides of the equation:

$$\log n = \log a + \log N + \log E$$

which demonstrates an additive relationship, on a logarithmic scale, between n, N and E although in the original equation ($n = aNE$) there is a multiplicative relationship between n, N and E.

Further descriptions of these important and fundamental mathematical functions are given by Grossman and Turner (1974), and their application in analyses such as analysis of variance and regression is described in Manly (1992).

References

Abdul Jalil, S. and Patterson, I.J. (1989). Effect of simulated goose grazing on yield of autumn-sown barley in north-east Scotland. *Journal of Applied Ecology* **26**, 897–912.

Advani, R. and Mathur, R.P. (1982). Experimental reduction of rodent damage to vegetable crops in Indian villages. *Agro-Ecosystems* **8**, 39–45.

Andelt, W.F. (1992). Effectiveness of livestock guarding dogs for reducing predation on domestic sheep. *Wildlife Society Bulletin* **20**, 55–62.

Andelt, W.F. and Burnham, K.P. (1993). Effectiveness of nylon lines for deterring rock doves from landing on ledges. *Wildlife Society Bulletin* **21**, 451–456.

Anderson, R.M. (1984). Strategies for the control of infectious diseases. In *Pest and Pathogen Control: Strategic, Tactical and Policy Models* (ed. G.R. Conway), pp. 109–141. John Wiley & Sons: Chichester.

Anderson, R.M. and May, R.M. (1979). Population biology of infectious diseases: Part I. *Nature* **280**, 361–367.

Anderson, R.M. and May, R.M. (1982). Directly transmitted infectious diseases: control by vaccination. *Science* **215**, 1053–1060.

Anderson, R.M. and May, R.M. (1991). *Infectious Diseases of Humans: Dynamics and Control*. Oxford University Press: Oxford.

Anderson, R.M., Jackson, H.C., May, R.M. and Smith, A.M. (1981). Population dynamics of fox rabies in Europe. *Nature* **289**, 765–771.

Andrewartha, H.G. and Birch, L.C. (1954). *The Distribution and Abundance of Animals*. University of Chicago Press: Chicago.

Andrzejewski, R. and Jezierski, W. (1978). Management of a wild boar population and its effects on commercial land. *Acta Theriologica* **23**, 309–339.

Arthur, A.D., Pech, R.P. and Dickman, C.R. (2004). Habitat structure mediates the non-lethal effects of predation on enclosed populations of house mice. *Journal of Applied Ecology* **73**, 867–877.

Avery, M.L., Eiselman, D.S., Young, M.K., Humphrey, J.S. and Decker, D.G. (1999). Wading bird predation at tropical aquaculture facilities in central Florida. *North American Journal of Aquaculture* **61**, 64–69.

Bailey, J.A. (1984). *Principles of Wildlife Management*. John Wiley & Sons: New York.

Banks, P.B. (1999). Predation by introduced foxes on native bush rats in Australia: do foxes take the doomed surplus? *Journal of Applied Ecology* **36**, 1063–1071.

Banks, P.B., Dickman, C.R. and Newsome, A.E. (1998). Ecological costs of feral predator control: foxes and rabbits. *Journal of Wildlife Management* **62**, 766–772.

Barlow, N.D. (1987). Pastures, pests and productivity: simple grazing models with two herbivores. *New Zealand Journal of Ecology* **10**, 43–55.

Barlow, N.D. (1991). A spatially aggregated disease/host model for bovine TB in New Zealand possum populations. *Journal of Applied Ecology* **28**, 777–793.

Barlow, N.D. (2000). Non-linear transmission and simple models for bovine tuberculosis. *Journal of Animal Ecology* **69**, 703–713.

Barlow, N.D., Kean, J.M. and Briggs, C.J. (1997). Modelling the relative efficacy of culling and sterilisation for controlling populations. *Wildlife Research* **24**, 129–141.

Barreto, G.R., Rushton, S.P., Strachan, R. and Macdonald, D.W. (1998). The role of habitat and mink predation in determining the status and distribution of water voles in England. *Animal Conservation* **1**, 129–137.

Bascompte, J. and Rodriguez-Trelles, F. (1998). Eradication thresholds in epidemiology, conservation biology and genetics. *Journal of Theoretical Biology* **192**, 415–418.

Bayliss, P. (1989). Population dynamics of magpie geese in relation to rainfall and density: implications for harvest models in a fluctuating environment. *Journal of Applied Ecology* **26**, 913–924.

Bayliss, P. and Choquenot, D. (2003). The numerical response: rate of increase and food limitation in herbivores and predators. In *Wildlife Population Growth Rates* (eds R.M. Sibly, J. Hone and T.H. Clutton-Brock), pp. 148–179. Cambridge University Press: Cambridge.

Bayliss, P. and Yeomans, K.M. (1989). Distribution and abundance of feral livestock in the 'Top End' of the Northern Territory (1985–86), and their relation to population control. *Australian Wildlife Research* **16**, 651–676.

Beale, C.M. and Monahan, P. (2004). Human disturbance: people as predation-free predators? *Journal of Applied Ecology* **41**, 335–343.

Beddington, J.R. and May, R.M. (1977). Harvesting natural populations in a randomly fluctuating environment. *Science* **197**, 463–465.

Begon, M., Harper, J.L. and Townsend, C.R. (1996). *Ecology: Individuals, Populations and Communities*, 3rd edn. Blackwell Scientific Publications: Oxford.

Begon, M., Bennett, M., Bowers, R.G., French, N.P., Hazel, S.M. and Turner, J. (2002). A clarification of transmission terms in host-microparasite models: numbers, densities and areas. *Epidemiology and Infection* **129**, 147–153.

Belsky, A.J. (1986). Does herbivory benefit plants? A review of the evidence. *American Naturalist* **127**, 870–892.

Berryman, A.A. (1999). *Principles of Population Dynamics and their Application*. Stanley Thornes: Cheltenham.

Bertram, D.F. (1995). The roles of introduced rats and commercial fishing in the decline of ancient murrelets on Langara Island, British Columbia. *Conservation Biology* **9**, 865–872.

Bicknell, K. (1993). Cost-benefit and cost-effectiveness analyses in pest management. *New Zealand Journal of Zoology* **20**, 307–312.

Blackburn, T.M., Cassey, P., Duncan, R.P., Evans, K.L. and Gaston, K.J. (2004). Avian extinction and mammalian introductions on oceanic islands. *Science* **305**, 1955–1958.

Blackburn, T.M., Petchey, O.L., Cassey, P. and Gaston, K.J. (2005). Functional diversity of mammalian predators and extinction in island birds. *Ecology* **86**, 2916–2923.

Blueweiss, L., Fox, H., Kudzma, V., Nakashima, D., Peters, R. and Sams, S. (1978). Relationships between body size and some life history parameters. *Oecologia* **37**, 257–272.

Bollinger, E.K. and Caslick, J.W. (1984). Relationships between bird-dropping counts, corn damage, and red-winged blackbird activity in field corn. *Journal of Wildlife Management* **48**, 209–211.

Bomford, M. and O'Brien, P. (1995). Eradication or control of vertebrate pests? *Wildlife Society Bulletin* **23**, 249–255.

Bomford, M., Newsome, A. and O'Brien, P. (1995). Solutions to feral animal problems: ecological and economic principles. In *Conserving Biodiversity: Threats and Solutions* (eds R.A. Bradstock, T.D. Auld, D.A. Keith, R.T. Kingsford, D. Lunney and D.P. Sivertsen), pp. 202–209. Surrey Beatty and Sons: Chipping Norton.

Boyle, S.A. and Samson, F.B. (1985). Effects of nonconsumptive recreation on wildlife: a review. *Wildlife Society Bulletin* **13**, 110–116.

Bradbury, R.B., Payne, R.J.H., Wilson, J.D. and Krebs, J.R. (2001). Predicting population responses to resource management. *Trends in Ecology and Evolution* **16**, 440–445.

Bratton, S.P. (1975). The effect of the European wild boar, *Sus scrofa*, on gray beech forest in the Great Smoky Mountains National Park. *Ecology* **56**, 1356–1366.

Braysher, M. (1993). *Managing Vertebrate Pests: Principles and Strategies.* Australian Government Publishing Service: Canberra.

Brown, P.R. (1993). Pasture response following rabbit control on grazing land. M.App.Sci. thesis. University of Canberra: Canberra.

Bruggers, R.L. and Elliott, C.C.H. (1989). *Quelea quelea: Africa's Bird Pest*. Oxford University Press: Oxford.

Bryce, J., Pritchard, J.S., Waran, N.K. and Young, R.J. (1997). Comparison of methods for obtaining population estimates for red squirrels in relation to damage due to bark stripping. *Mammal Review* **27**, 165–170.

Buckle, A.P., Rowe, F.P. and Husin, A.R. (1984). Field trials of warfarin and brodifacoum wax block baits for the control of the rice field rat, *Rattus argentiventer*, in peninsular Malaysia. *Tropical Pest Management* **30**, 51–58.

Burbidge, A.A. and McKenzie, N.L. (1989). Patterns in the modern decline of Western Australia's vertebrate fauna: causes and conservation implications. *Biological Conservation* **50**, 143–198.

Burger, J. (1985). Factors affecting bird strikes on aircraft at a coastal airport. *Biological Conservation* **33**, 1–28.

Burgman, M.A. and Lindenmayer, D.B. (1998). *Conservation Biology for the Australian Environment.* Surrey Beatty and Sons: Chipping Norton.

Burnham, K.P. and Anderson, D.R. (1998). *Model Selection Procedures: A Practical Information-Theoretic Approach.* Springer: New York.

Burnham, K.P. and Anderson, D.R. (2002). *Model Selection and Multimodel Inference: A Practical Information-Theoretic Approach,* 2nd edn. Springer: New York.

Burt, W.H. and Grossenheider, R.P. (1976). *A Field Guide to the Mammals,* 3rd edn. Houghton Mifflin: Boston.

Byrom, A. (2002). Dispersal and survival of juvenile feral ferrets *Mustelo furo* in New Zealand. *Journal of Applied Ecology* **39**, 67–78.

Caley, P. (1993). Population dynamics of feral pigs (*Sus scrofa*) in a tropical riverine habitat complex. *Wildlife Research* **20**, 625–636.

Caley, P. (1998). Broad-scale possum and ferret correlates of macroscopic *Mycobacterium bovis* infection in feral ferret populations. *New Zealand Veterinary Journal* **46**, 157–162.

Caley, P. (2001). Inference on the host status of feral ferrets (*Mustela furo*) in New Zealand for *Mycobacterium bovis* infection. PhD thesis. University of Canberra: Canberra.

Caley, P. and Hone, J. (2004). Disease transmission between and within species, and the implications for disease control. *Journal of Applied Ecology* **41**, 94–104.

Caley, P. and Ottley, B. (1995). The effectiveness of hunting dogs for removing feral pigs (*Sus scrofa*). *Wildlife Research* **22**, 147–154.

Caley, P., Hickling, G.J., Cowan, P.E. and Pfeiffer, D.U. (1999). Effects of sustained control of brushtail possums on levels of *Mycobacterium bovis* infection in cattle and brushtail possum populations from Hohotaka, New Zealand. *New Zealand Veterinary Journal* **47**, 133–142.

Caley, P., Hone, J. and Cowan, P.E. (2001). The relationship between prevalence of *Mycobacterium bovis* infection in feral ferrets and possum abundance. *New Zealand Veterinary Journal* **49**, 195–200.

Carpenter, S.R. (1981). Effect of control measures on pest populations subject to regulation by parasites and pathogens. *Journal of Theoretical Biology* **92**, 181–184.

Case, T.J. and Bolger, D.T. (1991). The role of introduced species in shaping the distribution and abundance of island reptiles. *Evolutionary Ecology* **5**, 272–290.

Caswell, H. (1978). Predator-mediated coexistence: a non-equilibrium model. *American Naturalist* **112**, 127–154.

Caughley, G. (1970). Eruption of ungulate populations, with emphasis on Himalayan thar in New Zealand. *Ecology* **51**, 53–72.

Caughley, G. (1980). *Analysis of Vertebrate Populations*. Reprinted with corrections. John Wiley & Sons: New York.

Caughley, G. (1987). Ecological relationships. In *Kangaroos. Their Ecology and Management in the Sheep Rangelands of Australia* (eds G. Caughley, N. Shepherd and J. Short), pp. 159–187. Cambridge University Press: Cambridge.

Caughley, G. (1994). Directions in conservation biology. *Journal of Animal Ecology* **63**, 215–244.

Caughley, G. and Gunn, A. (1996). *Conservation Biology in Theory and Practice*. Blackwell Scientific: Oxford.

Caughley, G. and Krebs, C.J. (1983). Are big mammals simply small mammals writ large? *Oecologia* **59**, 7–17.

Caughley, G. and Sinclair, A.R.E. (1994). *Wildlife Ecology and Management*. Blackwell Scientific: London.

Caughley, J., Bomford, M., Parker, B., Sinclair, R., Griffiths, J. and Kelly, D. (1998). *Managing Vertebrate Pests: Rodents*. Bureau of Resource Sciences: Canberra.

Chamberlin, T.C. (1965). The method of multiple working hypotheses. *Science* **148**, 754–759.

Chapuis, J.L., Bousses, P. and Barnaud, G. (1994). Alien mammals, impact and management in the French subantarctic islands. *Biological Conservation* **67**, 97–104.

Charnov, E.L. (1976). Optimal foraging: the marginal value theorem. *Theoretical Population Biology* **9**, 129–136.

Charnov, E.L. (1993). *Life History Invariants. Some Explorations of Symmetry in Evolutionary Ecology*. Oxford University Press: Oxford.

Cheeseman, C.L., Jones, G.W., Gallagher, J. and Mallinson, P.J. (1981). The population structure, density and prevalence of tuberculosis (*Mycobacterium bovis*) in badgers (*Meles meles*) from four areas in south-west England. *Journal of Applied Ecology* **18**, 795–804.

Cheeseman, C.L., Wilesmith, J.W. and Stuart, F.A. (1989). Tuberculosis: the disease and its epidemiology in the badger, a review. *Epidemiology and Infection* **103**, 113–125.

Choquenot, D. (1988). Feral donkeys in northern Australia: population dynamics and the cost of control. M.App.Sci. thesis. Canberra College of Advanced Education: Canberra.

Choquenot, D. (1991). Density-dependent growth, body condition, and demography in feral donkeys: testing the food hypothesis. *Ecology* **72**, 805–813.

Choquenot, D. (1998). Testing the relative influence of intrinsic and extrinsic variation in food availability on feral pig populations in Australia's rangelands. *Journal of Animal Ecology* **67**, 887–907.

Choquenot, D., Hone, J. and Saunders, G. (1999). Using predator-prey theory to evaluate helicopter shooting for feral pig control. *Wildlife Research* **26**, 251–261.

Choquenot, D., Kilgour, R.J. and Lukins, B.S. (1993). An evaluation of feral pig trapping. *Wildlife Research* **20**, 15–22.

Choquenot, D., Lukins, B. and Curran, G. (1997). Assessing lamb predation by feral pigs in Australia's semi-arid rangelands. *Journal of Applied Ecology* **34**, 1445–1454.

Choquenot, D., McIlroy, J. and Korn, T. (1996). *Managing Vertebrate Pests: Feral Pigs*. Bureau of Resource Sciences: Canberra.

Cilento, N.J. and Jones, D.N. (1999). Aggression by Australian magpies *Gymnorhina tibicen* towards human intruders. *Emu* **99**, 85–90.

Clark, C.W. (1976). *Mathematical Bioeconomics: The Optimal Management of Renewable Resources*. John Wiley & Sons: New York.

Clark, C.W. (1981). Bioeconomics. In *Theoretical Ecology: Principles and Applications*, 2nd edn (ed. R.M. May), pp. 387–418. Blackwell Scientific: Oxford.

Cleaveland, S. and Dye, C. (1995). Maintenance of a microparasite infecting several host species: rabies in the Serengeti. *Parasitology* **111**, S33–S47.

Cleaveland, S., Hess, G.R., Dobson, A.P., Laurenson, M.K., McCallum, H.I., Roberts, M.G. and Woodroffe, R. (2001). The role of pathogens in biological conservation. In *The Ecology of Wildlife Diseases* (eds P.J. Hudson, A. Rizzoli, B.T. Grenfell, H. Heesterbeek and A.P. Dobson), pp. 139–150. Oxford University Press: Oxford.

Clutton-Brock, T.H. (1988). *Reproductive Success*. University of Chicago Press: Chicago.

Clutton-Brock, T.H. and Albon, S. (1989). *Red Deer in the Highlands*. BSP Professional Books: Oxford.

Coleman, J.D. (1988). Distribution, prevalence, and epidemiology of bovine tuberculosis in brushtail possums, *Trichosurus vulpecula*, in the Hohonu range, New Zealand. *Australian Wildlife Research* **15**, 651–663.

Coleman, J.D. and Cooke, M.M. (2001). *Mycobacterium bovis* infection in wildlife in New Zealand. *Tuberculosis* **81**, 191–202.

Collins, A.R., Workman, J.P. and Uresk, D.W. (1984). An economic analysis of black-tailed prairie dog (*Cynomys ludovicianus*) control. *Journal of Range Management* **37**, 358–361.

Connell, J.H. (1978). Diversity in tropical rainforests and coral reefs. *Science* **199**, 1302–1310.

Conner, M.M., Jaeger, M.M., Weller, T.J. and McCullough, D.R. (1998). Effect of coyote removal on sheep depredation in northern California. *Journal of Wildlife Management* **62**, 690–699.

Conover, M.R. (1997a). Wildlife management by metropolitan residents in the United States: practices, perceptions, costs, and values. *Wildlife Society Bulletin* **25**, 306–311.

Conover, M.R. (1997b). Monetary and intangible valuation of deer in the United States. *Wildlife Society Bulletin* **25**, 298–305.

Conover, M. (2002). *Resolving Human–Wildlife Conflicts. The Science of Wildlife Damage Management*. CRC Press: Boca Raton.

Conover, M.R. and Decker, D.J. (1991). Wildlife damage to crops: perceptions of agricultural and wildlife professionals in 1957 and 1987. *Wildlife Society Bulletin* **19**, 46–52.

Conover, M.R., Pitt, W.C., Kessler, K.K., DuBow, T.J. and Sanborn, W.A. (1995). Review of human injuries, illnesses, and economic losses caused by wildlife in the United States. *Wildlife Society Bulletin* **23**, 407–414.

Cooke, B.D. (1981). Rabbit control and the conservation of native mallee vegetation on roadsides in South Australia. *Australian Wildlife Research* **8**, 627–636.

Cooray, R.G. and Mueller-Dombois, D. (1981). Feral pig activity. In *Island Ecosystems: Biological Organisation in Selected Hawaiian Communities* (eds D. Mueller-Dombois, K.W. Bridges and H.L.

Carson), pp. 309–319. Hutchinson Research Publishing Co.: Stroudsburg.

Copson, G. and Whinam, J. (1998). Response of vegetation on subantarctic Macquarie Island to reduced rabbit grazing. *Australian Journal of Botany* **46**, 15–24.

Corbett, L. (1995). Does dingo predation or buffalo competition regulate feral pig populations in the Australian wet-dry tropics? An experimental study. *Wildlife Research* **22**, 65–74.

Cote, I.M. and Sutherland, W.J. (1997). The effectiveness of removing predators to protect bird populations. *Conservation Biology* **11**, 395–405.

Coulson, G.M. (1982). Road-kills of macropods on a section of highway in central Victoria. *Australian Wildlife Research* **9**, 21–26.

Courchamp, F., Grenfell, B. and Clutton-Brock, T. (1999a). Population dynamics of obligate cooperators. *Proceedings of the Royal Society London B* **266**, 557–563.

Courchamp, F., Langlais, M. and Sugihara, G. (1999b). Control of rabbits to protect birds from cat predation. *Biological Conservation* **89**, 219–225.

Courchamp, F., Langlais, M. and Sugihara, G. (1999c). Cats protecting birds: modelling the mesopredator release effect. *Journal of Animal Ecology* **68**, 282–292.

Crabb, A.C., Salmon, T.P. and Marsh, R.E. (1986). Bird problems in California pistachio production. In *Proceedings of the Twelfth Vertebrate Pest Conference* (ed. T.P. Salmon), pp. 295–302. University of California: Davis.

Croft, D. (1990). The impact of rabbits on sheep production. M.Sc. thesis. University of New South Wales: Sydney.

Croft, J.D., Fleming, P.J.S. and van de Ven, R. (2002). The impact of rabbits on a grazing system in eastern New South Wales. 1. Ground cover and pastures. *Australian Journal of Experimental Agriculture* **42**, 909–916.

Cullen, R., Hughey, K.F.D., Fairburn, G. and Moran, E. (2005). Economic analyses to aid nature conservation decision making. *Oryx* **39**, 327–334.

Cummings, J.L., Guarino, J.L. and Knittle, C.E. (1989). Chronology of blackbird damage to sunflowers. *Wildlife Society Bulletin* **17**, 50–52.

Daszak, P., Cunningham, A.A. and Hyatt, A.D. (2000). Emerging infectious diseases of wildlife – threats to biodiversity and human health. *Science* **287**, 443–449.

Davis, S.A., Begon, M., De Bruyn, L., Ageyev, V.S., Klassovskiy, N.L., Pole, S.B., Viljugrein, H., Stenseth, N.C. and Leirs, H. (2004a). Predictive thresholds for plague in Kazakhstan. *Science* **304**, 736–738.

Davis, S.A., Leirs, H., Pech, R.P., Zhang, Z. and Stenseth, N.C. (2004b). On the economic benefit of predicting rodent outbreaks in agricultural systems. *Crop Protection* **23**, 305–314.

Davis, S.A., Pech, R.P. and Catchpole, E.A. (2003). Populations in variable environments: the effect of variability in a species' primary resource. In *Wildlife Population Growth Rates* (eds R.M. Sibly, J. Hone and T.H. Clutton-Brock), pp. 180–197. Cambridge University Press: Cambridge.

Dennis, B. (2002). Allee effects in stochastic populations. *Oikos* **96**, 389–401.

Derrick, J. and Dann, P. (1997). Soils and agriculture. In *Human Ecology, Human Economy* (eds M. Diesendorf and C. Hamilton), pp. 171–196. Allen & Unwin: Sydney.

Dexter, N. (2003). Stochastic models of foot and mouth disease in feral pigs in the Australian semi-arid rangelands. *Journal of Applied Ecology* **40**, 293–306.

Diamond, J. (1985). Rats as agents of extinction. *Nature* **318**, 602–603.

Dietz, K. (1993). The estimation of the basic reproductive number for infectious diseases. *Statistical Methods in Medical Research* **2**, 23–41.

Dobbie, W.R., Berman, D.McK. and Braysher, M. (1993). *Managing Vertebrate Pests: Feral Horses*. Bureau of Resource Sciences: Canberra.

Dobson, A.P. and May, R.M. (1991). Parasites, cuckoos, and avian population dynamics. In *Bird Population Studies* (eds C.M. Perrins, J.-D. Lebreton and G.J.M. Hirons), pp. 391–412. Oxford University Press: Oxford.

Dobson, A. and Meagher, M. (1996). The population dynamics of brucellosis in the Yellowstone National Park. *Ecology* **77**, 1026–1036.

Dolbeer, R.A. (1999). Overview and management of vertebrate pests. In *Handbook of Pest Management* (ed. J.R. Ruberson), pp. 663–691. Marcel Dekker: New York.

Dolbeer, R.A., Holler, N.R. and Hawthorne, D.W. (1994). Identification and control of wildlife damage. In *Research and Management Techniques for Wildlife and Habitats*, 5th edn (ed. T.A. Bookhout), pp. 474–506. Wildlife Society: Bethesda.

Donlan, C.J., Tershy, B.R. and Croll, D.A. (2002). Islands and introduced herbivores: conservation action as ecosystem experimentation. *Journal of Applied Ecology* **39**, 235–246.

Donnelly, C.A., Woodroffe, R., Cox, D.R., Bourne, J., Gettinby, G., Le Fevre, A.M., McInerney, J.P. and Morrison, W.I. (2003). Impact of localized badger culling on tuberculosis incidence in British cattle. *Nature* **426**, 834–837.

Donnelly, C.A., Woodroffe, R., Cox, D.R., Bourne, F.J., Cheeseman, C.L., Clifton-Hadley, R.S., Wei, G., Gettinby, G., Gilks, P., Jenkins, H., Johnston, W.T., Le Fevre, A.M., McInerney, J.P. and Morrison, W.I. (2006). Positive and negative effects of widespread badger culling on tuberculosis in cattle. *Nature* **439**, 843–846.

Dudley, S.F.J., Haestier, R.C., Cox, K.R. and Murray, M. (1998). Shark control: experimental fishing with baited drumlines. *Marine and Freshwater Research* **49**, 653–661.

Duncan, R.P., Bomford, M., Forsyth, D.M. and Conibear, L. (2001). High predictability in introduction outcomes and the geographical range size of introduced Australian birds: a role for climate. *Journal of Animal Ecology* **70**, 621–632.

Dyring, J. (1990). The impact of feral horses (*Equus caballus*) on subalpine and montane environments in Australia. M.App.Sci. thesis. University of Canberra: Canberra.

Eberhardt, L.L. (1985). Assessing the dynamics of wild populations. *Journal of Wildlife Management* **49**, 997–1012.

Eberhardt, L.L. (1987). Population projections from simple models. *Journal of Applied Ecology* **24**, 103–118.

Eberhardt, L.L. (2002). A paradigm for population analysis of long-lived vertebrates. *Ecology* **83**, 2841–2854.

Efford, M., Warburton, B. and Spencer, N. (2000). Home-range changes by brushtail possums in response to control. *Wildlife Research* **27**, 117–127.

Elton, C. (1931). The study of epidemic diseases among wild animals. *Journal of Hygiene* **31**, 435–456.

Empson, R.A. and Miskelly, C.M. (1999). The risks, costs and benefits of using brodifacoum to eradicate rats from Kapiti Island, New Zealand. *New Zealand Journal of Ecology* **23**, 241–254.

Engeman, R.M., Shwiff, S.A., Constantin, B., Stahl, M. and Smith, H. T. (2002). An economic analysis of predator removal approaches for protecting marine turtle nests at Hobe Sound National Wildlife Refuge. *Ecological Economics* **42**, 469–478.

Engeman, R.M., Shwiff, S.A., Cano, F. and Constantin, B. (2003). An economic assessment of the potential for predator management to benefit Puerto Rican parrots. *Ecological Economics* **46**, 283–292.

Engeman, R.M., Vice, D.S., Nelson, G. and Muna, E. (2000). Brown tree snake effectively removed from a large plot of land on Guam by perimeter trapping. *International Biodeterioration & Biodegradation* **45**, 139–142.

Fan, N., Zhou, W., Wei, W., Wang, Q. and Jiang, Y. (1999). Rodent pest management in the Qinghai-Tibet alpine meadow ecosystem. In *Ecologically-based Management of Rodent Pests* (eds G.R. Singleton, L.A. Hinds, H. Leirs and Z. Zhang), pp. 285–304. Australian Centre for International Agricultural Research: Canberra.

Feare, C.J. (1974). Ecological studies of the rook (*Corvus frugilegus* L.) in north-east Scotland. Damage and its control. *Journal of Applied Ecology* **11**, 897–914.

Feare, C.J. (1991). Control of bird pest populations. In *Bird Population Studies* (eds C.M. Perrins, J.-D. Lebreton and G.J.M. Hirons), pp. 463–478. Oxford University Press: Oxford.

Fenner, F. and Fantini, B. (1999). *Biological Control of Vertebrate Pests. The History of Myxomatosis, an Experiment in Evolution.* CABI Publishing: New York.

Fenton, A., Fairbairn, J.P., Norman, R. and Hudson, P.J. (2002). Parasite transmission: reconciling theory and reality. *Journal of Animal Ecology* **71**, 893–905.

Ferreras, P. and Macdonald, D.W. (1999). The impact of American mink *Mustela vison* on water birds in the upper Thames. *Journal of Applied Ecology* **36**, 701–708.

Fleming, P.J.S., Croft, J.D. and Nicol, H.I. (2002). The impact of rabbits on a grazing system in eastern New South Wales. 2. Sheep production. *Australian Journal of Experimental Agriculture* **42**, 917–923.

Foran, B., Low, W. and Strong, B. (1985). The response of rabbit populations and vegetation to rabbit control on a calcareous shrubby grassland in central Australia. *Australian Wildlife Research* **12**, 237–247.

Forsyth, D.M. and Caley, P. (2006). Testing the irruptive paradigm of large-herbivore dynamics. *Ecology* **87**, 297–303.

Forsyth, D.M. and Duncan, R.P. (2001). Propagule size and the relative success of exotic ungulate and bird introductions to New Zealand. *American Naturalist* **157**, 583–595.

Forsyth, D.M., Parkes, J.P. and Hickling, G.J. (2000). A case for multi-species management of sympatric herbivore pest impacts in the central Southern Alps, New Zealand. *New Zealand Journal of Ecology* **24**, 97–103.

Forsyth, D.M., Hone, J., Parkes, J.P., Reid, G.H. and Stronge, D. (2003). Feral goat control in Egmont National Park, New Zealand, and the implications for eradication. *Wildlife Research* **30**, 437–450.

Friend, J.A. and Thomas, N.D. (2003). Conservation of the numbat (*Myrmecobius fasciatus*). In *Predators with Pouches* (eds M. Jones, C. Dickman and M. Archer), pp. 452–463. CSIRO Publishing: Melbourne.

Fritts, T.H. and Rodda, G.H. (1998). The role of introduced species in the degradation of island ecosystems: a case study of Guam. *Annual Review of Ecology and Systematics* **29**, 113–140.

Giles, J.R. (1980). Ecology of feral pigs in New South Wales. PhD thesis. University of Sydney: Sydney.

Gill, J.E., Kerins, G.M., Langton, S.D. and MacNicoll, A.D. (1994). Blood-clotting response time test for bromadiolone resistance in Norway rats. *Journal of Wildlife Management* **58**, 454–461.

Gillespie, G.D. (1985). Feeding behaviour and impact of ducks on ripening barley crops grown in Otago, New Zealand. *Journal of Applied Ecology* **22**, 347–356.

Gilpin, M.E. and Soule, M.E. (1986). Minimum viable populations: processes of species extinction. In *Conservation Biology* (ed. M.E. Soule), pp. 19–34. Sinauer Associates: Sunderland.

Glahn, J.F., Rasmussen, E.S., Tomsa, T. and Preusser, K.J. (1999). Distribution and relative impact of avian predators at aquaculture facilities in the northeastern United States. *North American Journal of Aquaculture* **61**, 340–348.

Goodloe, R.B., Warren, R.J., Cothran, E.G., Bratton, S.P. and Trembicki, K.A. (1991). Genetic variation and its management applications in eastern U.S. feral horses. *Journal of Wildlife Management* **55**, 412–421.

Goodrich, J.M. and Buskirk, S.W. (1995). Control of abundant native vertebrates for conservation of endangered species. *Conservation Biology* **9**, 1357–1364.

Gorynska, W. (1981). Method of determining relations between the extent of damage in farm crops, big game numbers, and environmental conditions. *Acta Theriologica* **26**, 469–481.

Gosling, L.M. and Baker, S.J. (1989). The eradication of muskrats and coypus from Britain. *Biological Journal of the Linnean Society* **38**, 39–51.

Greaves, J.H. (1985). The present status of resistance to anticoagulants. *Acta Zoologica Fennica* **173**, 159–162.

Green, R.E. and Etheridge, B. (1999). Breeding success of the hen harrier *Circus cyaneus* in relation to the distribution of grouse moors and the red fox *Vulpes vulpes*. *Journal of Applied Ecology* **36**, 472–483.

Green, R.J. and Higginbottom, K. (2000). The effects of non-consumptive wildlife tourism on free-ranging wildlife: a review. *Pacific Conservation Biology* **6**, 183–197.

Greentree, C., Saunders, G., McLeod, L. and Hone, J. (2000). Lamb predation and fox control in south-eastern Australia. *Journal of Applied Ecology* **37**, 935–943.

Grenfell, B.T. and Dobson, A.P. (1995). *Ecology of Infectious Diseases in Natural Populations*. Cambridge University Press: Cambridge.

Grenfell, B. and Harwood, J. (1997). (Meta)population dynamics of infectious diseases. *Trends in Ecology and Evolution* **12**, 395–399.

Gribble, N.A., McPherson, G. and Lane, B. (1998). Effect of the Queensland shark control program on non-target species: whale, dugong, turtle and dolphin: a review. *Marine and Freshwater Research* **49**, 645–651.

Grime, J.P. (1973). Competitive exclusion in herbaceous vegetation. *Nature* **242**, 344–347.

Groot Bruinderink, G.W.T.A. and Hazebroek, E. (1996). Ungulate traffic collisions in Europe. *Conservation Biology* **10**, 1059–1067.

Grossman, S.I. and Turner, J.E. (1974). *Mathematics for the Biological Sciences*. Macmillan: New York.

Gupta, S., Trenholme, K., Anderson, R.M. and Day, K.P. (1994). Antigenic diversity and the transmission dynamics of *Plasmodium falciparum*. *Science* **263**, 961–963.

Harding, E.K., Doak, D.F. and Albertson, J.D. (2001). Evaluating the effectiveness of predator control: the non-native red fox as a case study. *Conservation Biology* **15**, 1114–1122.

Harper, J.L. and Hawksworth, D.L. (1994). Biodiversity: measurement and estimation. *Philosophical Transactions of the Royal Society London B* **345**, 5–12.

Headley, J.C. (1972). Defining the economic threshold. In *Pest Control Strategies for the Future*, pp. 100–108. National Research Council: Washington.

Hickling, G.J., Henderson, R.J. and Thomas, M.C.C. (1999). Poisoning mammalian pests can have unintended consequences for future control: two case studies. *New Zealand Journal of Ecology* **23**, 267–273.

Hik, D.S. (1995). Does risk of predation influence population dynamics? Evidence from the cyclic decline of snowshoe hares. *Wildlife Research* **22**, 115–129.

Hoare, R.E. (1999). Determinants of human–elephant conflict in a land-use mosaic. *Journal of Applied Ecology* **36**, 689–700.

Hobbs, N.T., Bowden, D.C. and Baker, D.L. (2000). Effects of fertility control on populations of ungulates: general, stage-structured models. *Journal of Wildlife Management* **64**, 473–491.

Hobson, K.A., Drever, M.C. and Kaiser, G.W. (1999). Norway rats as predators of burrow-nesting seabirds: insights from stable isotope analyses. *Journal of Wildlife Management* **63**, 14–25.

Hone, J. (1983). A short-term evaluation of feral pig eradication at Willandra in western New South Wales. *Australian Wildlife Research* **10**, 269–275.

Hone, J. (1988a). Evaluation of methods for ground survey of feral pigs and their sign. *Acta Theriologica* **33**, 451–465.

Hone, J. (1988b). Feral pig rooting in a mountain forest and woodland: distribution, abundance and relationships with environmental variables. *Australian Journal of Ecology* **13**, 393–400.

Hone, J. (1990). Predator-prey theory and feral pig control, with emphasis on evaluation of shooting from a helicopter. *Australian Wildlife Research* **17**, 123–130.

Hone, J. (1992). Modelling of poisoning for vertebrate pest control, with emphasis on poisoning feral pigs. *Ecological Modelling* **62**, 311–327.

Hone, J. (1994a). *Analysis of Vertebrate Pest Control*. Cambridge University Press: Cambridge.

Hone, J. (1994b). A mathematical model of detection and dynamics of porcine transmissible gastroenteritis. *Epidemiology and Infection* **113**, 187–197.

Hone, J. (1995). Spatial and temporal aspects of vertebrate pest damage, with emphasis on feral pigs. *Journal of Applied Ecology* **32**, 311–319.

Hone, J. (1996). Analysis of vertebrate pest research. In *Proceedings of the Seventeenth Vertebrate Pest Conference* (eds R.M. Timm and A.C. Crabb), pp. 13–17. University of California: Davis.

Hone, J. (1999). On rate of increase (*r*): patterns of variation in Australian mammals and the implications for wildlife management. *Journal of Applied Ecology* **36**, 709–718.

Hone, J. (2002). Feral pigs in Namadgi National Park, Australia: dynamics, impacts and management. *Biological Conservation* **105**, 231–242.

Hone, J. (2004). Yield, compensation and fertility control: a model for vertebrate pests. *Wildlife Research* **31**, 357–368.

Hone, J. (2006). Linking pasture, livestock productivity and vertebrate pest management. *New Zealand Journal of Ecology* **30**, 13–23.

Hone, J. and Stone, C.P. (1989). A comparison and evaluation of feral pig management in two national parks. *Wildlife Society Bulletin* **17**, 419–425.

Hone, J., Pech, R. and Yip, P. (1992). Estimation of the dynamics and rate of transmission of classical swine fever (hog cholera) in wild pigs. *Epidemiology and Infection* **108**, 377–386.

Hubbard, M.W., Danielson, B.J. and Schmitz, R.A. (2000). Factors influencing the location of deer–vehicle accidents in Iowa. *Journal of Wildlife Management* **64**, 707–713.

Hudson, P.J., Rizzoli, A., Grenfell, B.T., Heesterbeek, H. and Dobson, A.P. (2001). *The Ecology of Wildlife Diseases*. Oxford University Press: Oxford.

Innes, J., Hay, R., Flux, I., Bradfield, P., Speed, H. and Jansen, P. (1999). Successful recovery of North Island kokako (*Callaeas cinerea wilsoni*) populations, by adaptive management. *Biological Conservation* **87**, 201–214.

Izac, A.-M. and O'Brien, P. (1991). Conflict, uncertainty and risk in feral pig management: the Australian approach. *Journal of Environmental Management* **32**, 1–18.

Jackson, D.B. (2001). Experimental removal of introduced hedgehogs improves wader nest success in the Western Isles, Scotland. *Journal of Applied Ecology* **38**, 802–812.

Jacob, J., Herawati, N.A., Davis, S.A., Singleton, G.R. (2004). The impact of sterilized females on enclosed populations of ricefield rats. *Journal of Wildife Management* **68**, 1130–1137.

Johnson, D.B., Guthery, F.S. and Koerth, N.E. (1989). Grackle damage to grapefruit in the lower Rio Grande valley. *Wildlife Society Bulletin* **17**, 46–50.

Jones, D. (2002). *Magpie Alert. Learning to Live with a Wild Neighbour.* University of New South Wales Press: Sydney.

Jones, D.N. and Thomas, L.K. (1999). Attacks on humans by Australian magpies: management of an extreme suburban human–wildlife conflict. *Wildlife Society Bulletin* **27**, 473–478.

Kadmon, R. and Tielborger, K. (1999). Testing for source-sink population dynamics: an experimental approach exemplified with desert annuals. *Oikos* **86**, 417–429.

Kallen, A., Arcuri, P. and Murray, J.D. (1985). A simple model for the spatial spread and control of rabies. *Journal of Theoretical Biology* **116**, 377–393.

Katahira, L.K., Finnegan, P. and Stone, C.P. (1993). Eradicating feral pigs in montane mesic habitat at Hawaii Volcanoes National Park. *Wildlife Society Bulletin* **21**, 269–274.

Kean, J.M., Barlow, N.D. and Hickling, G.J. (1999). Evaluating potential sources of bovine tuberculosis infection in a New Zealand cattle herd. *New Zealand Journal of Agricultural Research* **42**, 101–106.

Keeling, M.J. and Gilligan, C.A. (2000). Metapopulation dynamics of bubonic plague. *Nature* **407**, 903–906.

Kenward, R.E., Hall, D.G., Walls, S.S. and Hodder, K.H. (2001). Factors affecting predation by buzzards *Buteo buteo* on released pheasants *Phasianus colchicus*. *Journal of Applied Ecology* **38**, 813–822.

Kerr, W.A. and Wilman, E.A. (1988). A proposed formal structure for assessing bear–human encounters. *Environmental Management* **12**, 173–179.

Kessler, W.B., Csanyi, S. and Field, R. (1998). International trends in university education for wildlife conservation and management. *Wildlife Society Bulletin* **26**, 927–936.

King, D.R., Twigg, L.E. and Gardner, J.L. (1989). Tolerance to sodium monofluoroacetate in dasyurids from Western Australia. *Australian Wildlife Research* **16**, 131–140.

Kinnear, J.E., Onus, M.L. and Sumner, N.R. (1998). Fox control and rock-wallaby population dynamics – II. An update. *Wildlife Research* **25**, 81–88.

Klocker, U., Croft, D.B. and Ramp, D. (2006). Frequency and causes of kangaroo–vehicle collisions on an Australian outback highway. *Wildlife Research* **33**, 5–15.

Kokko, H. (2001). Optimal and suboptimal use of compensatory responses to harvesting: timing of hunting as an example. *Wildlife Biology* **7**, 141–150.

Komers, P.E. and Curman, G.P. (2000). The effect of demographic characteristics on the success of ungulate re-introductions. *Biological Conservation* **93**, 187–193.

Kraus, F., Campbell, E.W., Allison, A. and Pratt, T. (1999). *Eleutherodactylus* frog introductions to Hawaii. *Herpetological Review* **30**, 21–25.

Krebs, C.J. (1988). The experimental approach to rodent population dynamics. *Oikos* **52**, 143–149.

Krebs, C.J. (1999). *Ecological Methodology*, 2nd edn. Addison Wesley Longman: New York.

Krebs, C.J. (2001). *Ecology. The Experimental Analysis of Distribution and Abundance*, 5th edn. HarperCollins College Publishers: New York.

Kuhn, T.S. (1970). *The Structure of Scientific Revolutions*. University of Chicago Press: Chicago.

Lancia, R.A., Braun, C.E., Collopy, M.W., Dueser, R.D., Kie, J.G., Martinka, C.J., Nichols, J.D., Nudds, T.D., Porath, W.R. and Tilghman, N.G. (1996). ARM! For the future: adaptive resource management in the wildlife profession. *Wildlife Society Bulletin* **24**, 436–442.

Landa, A., Gudvangen, K., Swenson, J.E. and Roskaft, E. (1999). Factors associated with wolverine *Gulo gulo* predation on domestic sheep. *Journal of Applied Ecology* **36**, 963–973.

Lande, R. (1987). Extinction thresholds in demographic models of territorial populations. *American Naturalist* **130**, 624–635.

Lande, R. (1991). Population dynamics and extinction in heterogeneous environments: the northern spotted owl. In *Bird Population Studies* (eds C.M. Perrins, J.-D. Lebreton and G.J.M. Hirons), pp. 566–580. Oxford University Press: Oxford.

Lande, R., Engen, S. and Saether, B.-E. (1994). Optimal harvesting, economic discounting and extinction risk in fluctuating populations. *Nature* **372**, 88–90.

Lande, R., Saether, B.-E. and Engen, S. (1997). Threshold harvesting for sustainability of fluctuating resources. *Ecology* **78**, 1341–1350.

Lefebvre, L.W., Engeman, R.M., Decker, D.G. and Holler, N.R. (1989). Relationship of roof rat population indices with damage to sugarcane. *Wildlife Society Bulletin* **17**, 41–45.

Leirs, H., Verhagen, R., Verheyen, W., Mwanjabe, P. and Mbise, T. (1996). Forecasting rodent outbreaks in Africa: an ecological basis for *Mastomys* control in Tanzania. *Journal of Applied Ecology* **33**, 937–943.

Leopold, A. (1933). *Game Management*. University of Wisconsin Press: Madison.

Levin, P.S., Ellis, J., Petrik, R. and Hay, M.E. (2002). Indirect effects of feral horses on estuarine communities. *Conservation Biology* **16**, 1364–1371.

Levins, R. (1969). Some demographic and genetic considerations of environmental heterogeneity for biological control. *Bulletin of the Entomological Society of America* **15**, 237–240.

Liddle, M. (1997). *Recreation Ecology. The Ecological Impact of Outdoor Recreation and Ecotourism*. Chapman & Hall: London.

Linnell, J.D.C., Odden, J., Smith, M.E., Aanes, R. and Swenson, J.E. (1999). Large carnivores that kill livestock: do 'problem individuals' really exist? *Wildlife Society Bulletin* **27**, 698–705.

Lovell, C.D. and Dolbeer, R.A. (1999). Validation of the United States Air Force bird avoidance model. *Wildlife Society Bulletin* **27**, 167–171.

Maas, S. (1997). Population dynamics and control of feral goats in a semi-arid environment. M.App.Sci. thesis. University of Canberra: Canberra.

Maas, S. (1998). Feral goats in Australia: impacts and cost of control. In *Proceedings of the Eighteenth Vertebrate Pest Conference* (eds R.O. Baker and A.C. Crabb), pp. 100–106. University of California: Davis.

MacArthur, R.H. and Wilson, E.O. (1967). *The Theory of Island Biogeography*. Princeton University Press: Princeton.

Mackey, R.L. and Currie, D.J. (2001). The diversity-disturbance relationship: is it generally strong and peaked? *Ecology* **82**, 3479–3492.

Mackin, R. (1970). Dynamics of damage caused by wild boar to different agricultural crops. *Acta Theriologica* **27**, 447–458.

Macnab, J. (1983). Wildlife management as scientific experimentation. *Wildlife Society Bulletin* **11**, 397–401.

Malcolm, B., Sale, P. and Egan, A. (1996). *Agriculture in Australia. An Introduction*. Oxford University Press: Oxford.

Manly, B.F.J. (1992). *The Design and Analysis of Research Studies*. Cambridge University Press: Cambridge.

Manly, B.F.J. (2001). *Statistics for Environmental Science and Management*. Chapman & Hall/CRC: London.

Marshal, J.P. and Boutin, S. (1999). Power analysis of wolf–moose functional responses. *Journal of Wildlife Management* **63**, 396–402.

Martinsen, G.D., Cushman, J.H. and Whitham, T.G. (1990). Impact of pocket gopher disturbance on plant species diversity in a shortgrass prairie community. *Oecologia* **83**, 132–138.

May, R.M. (1976). Harvesting whale and fish populations. *Nature* **263**, 91–92.

May, R.M. and Anderson, R.M. (1979). Population biology of infectious diseases: Part II. *Nature* **280**, 455–461.

McArdle, B.H. (1996). Levels of evidence in studies of competition, predation and disease. *New Zealand Journal of Ecology* **20**, 7–15.

McCaffery, K.R. (1973). Road-kills show trends in Wisconsin deer populations. *Journal of Wildlife Management* **37**, 212–216.

McCallum, H. (1995). Modelling wildlife–parasite interactions to help plan and interpret field studies. *Wildlife Research* **22**, 21–29.

McCallum, H., Barlow, N. and Hone, J. (2001). How should pathogen transmission be modelled? *Trends in Ecology and Evolution* **16**, 295–300.

McIlroy, J.C. (1986). The sensitivity of Australian animals to 1080 poison. IX. Comparisons between the major groups of animals, and the potential danger non-target species face from 1080-poisoning campaigns. *Australian Wildlife Research* **13**, 39–48.

McIlroy, J.C. and Gifford, E.J. (1991). Effects on non-target animal populations of a rabbit trail-baiting campaign with 1080 poison. *Wildlife Research* **18**, 315–325.

McKay, H.V., Bishop, J.D., Feare, C.J. and Stevens, M.C. (1993). Feeding by brent geese can reduce yield of oilseed rape. *Crop Protection* **12**, 101–105.

McLeod, S.R. and Saunders, G.R. (2001). Improving management strategies for the red fox by using projection matrix analysis. *Wildlife Research* **28**, 333–340.

Meriggi, A. and Lovari, S. (1996). A review of wolf predation in southern Europe: does the wolf prefer wild prey to livestock? *Journal of Applied Ecology* **33**, 1561–1571.

Miller, B. and Mullette, K.J. (1985). Rehabilitation of an endangered Australian bird: the Lord Howe island woodhen *Tricholimnas sylvestris*. *Biological Conservation* **34**, 55–95.

Miller, B., Ceballos, G. and Reading, R. (1994). The prairie dog and biotic diversity. *Conservation Biology* **8**, 677–681.

Miller, K.K. and Jones, D.N. (2005). Wildlife management in Australasia: perceptions of objectives and priorities. *Wildlife Research* **32**, 265–272.

Miller, K.K. and Jones, D.N. (2006). Gender differences in the perceptions of wildlife management objectives and priorities in Australasia. *Wildlife Research* **33**, 155–159.

Milner-Gulland, E.J., Shea, K., Possingham, H., Coulson, T. and Wilcox, C. (2001). Competing harvesting strategies in a simulated population under uncertainty. *Animal Conservation* **4**, 157–167.

Mishra, C. (1997). Livestock depredation by large carnivores in the Indian trans-Himalaya: conflict perceptions and conservation prospects. *Environmental Conservation* **24**, 338–343.

Molsher, R. (1999). The ecology of feral cats, *Felis catus*, in open forest in New South Wales: interactions with food resources and foxes. PhD thesis. University of Sydney: Sydney.

Montague, T. (2000). *The Brushtail Possum. Biology, Impact and Management of an Introduced Marsupial.* Manaaki Whenua Press: Lincoln.

Moore, N., Whiterow, A., Kelly, P., Garthwaite, D., Bishop, J., Langton, S. and Cheeseman, C. (1999). Survey of badger *Meles meles* damage to agriculture in England and Wales. *Journal of Applied Ecology* **36**, 974–988.

Morgan, D.R., Milne, L., O'Connor, C. and Ruscoe, W.A. (2001). Bait shyness in possums induced by sublethal doses of cyanide paste bait. *International Journal of Pest Management* **47**, 277–284.

Morris, K., Johnson, B., Orell, P., Gaikhorst, G., Wayne, A. and Moro, D. (2003). Recovery of the threatened chuditch (*Dasyurus geoffroii*): a case study. In *Predators with Pouches* (eds M. Jones, C. Dickman, and M. Archer), pp. 435–451. CSIRO Publishing: Melbourne.

Morriss, G.A., Warburton, B. and Ruscoe, W.A. (2000). Comparison of the capture efficiency of a kill-trap set for brushtail possums that excludes ground-birds, and ground set leg-hold traps. *New Zealand Journal of Zoology* **27**, 201–206.

Murdoch, W.W. and Briggs, C.J. (1996). Theory for biological control: recent developments. *Ecology* **77**, 2001–2013.

Murdoch, W.W., Chesson, J. and Chesson, P.L. (1985). Biological control in theory and practice. *American Naturalist* **125**, 344–366.

Murphy, E.C., Robbins, L., Young, J.B. and Dowding, J.E. (1999). Secondary poisoning of stoats after an aerial 1080 poison operation in Pureora Forest, New Zealand. *New Zealand Journal of Ecology* **23**, 175–182.

Murua, R. and Rodriguez, J. (1989). An integrated control system for rodents in pine plantations in central Chile. *Journal of Applied Ecology* **26**, 81–88.

Mutze, G.J. (1993). Cost-effectiveness of poison bait trails for control of house mice in mallee cereal crops. *Wildlife Research* **20**, 445–456.

Naughton-Treves, L. (1998). Predicting patterns of crop damage by wildlife around Kibale National Park, Uganda. *Conservation Biology* **12**, 156–168.

Naughton-Treves, L., Treves, A., Chapman, C. and Wrangham, R. (1998). Temporal patterns of crop-raiding by primates: linking food availability in croplands and adjacent forest. *Journal of Applied Ecology* **35**, 596–606.

Newton, I. (1989). *Lifetime Reproduction in Birds*. Academic Press: London.

Newton, I. (1995). The contribution of some recent research on birds to ecological understanding. *Journal of Animal Ecology* **64**, 675–696.

Newton, I. (1998). *Population Limitation in Birds*. Academic Press: London.

Nichols, J.D. (1991). Responses of North American duck populations to exploitation. In *Bird Population Studies* (eds C.M. Perrins, J.-D. Lebreton and G.J.M. Hirons), pp. 498–525. Oxford University Press: Oxford.

Nichols, J.D., Conroy, D.R., Anderson, D.R. and Burnham, K.P. (1984). Compensatory mortality in waterfowl populations: a review of the evidence and implications for research and management. *Transactions of the North American Wildlife and Natural Resources Conference* **49**, 535–554.

Nielsen, C.K., Porter, W.F. and Underwood, H.B. (1997). An adaptive management approach to controlling suburban deer. *Wildlife Society Bulletin* **25**, 470–477.

Norbury, G.L., Norbury, D.C. and Heyward, R.P. (1998). Space use and denning behaviour of wild ferrets (*Mustela furo*) and cats (*Felis catus*). *New Zealand Journal of Ecology* **22**, 149–159.

Nugent, G., Fraser, W. and Sweetapple, P. (2001). Top down or bottom-up? Comparing the impacts of introduced arboreal possums and 'terrestrial' ruminants on native forests in New Zealand. *Biological Conservation* **99**, 65–79.

Oksanen, L. and Oksanen, T. (2000). The logic and realism of the hypothesis of exploitation ecosystems. *American Naturalist* **155**, 703–723.

Oksanen, L., Fretwell, S.D., Arruda, J. and Niemela, P. (1981). Exploitation ecosystems in gradients of primary productivity. *American Naturalist* **118**, 240–261.

Olsen, P. (1998). *Australia's Pest Animals: New Solutions to Old Problems.* Bureau of Resource Sciences/Kangaroo Press: Canberra.

O'Mairtin, D., Williams, D.H., Griffin, J.M., Dolan, L.A. and Eves, J.A. (1998). The effect of a badger removal programme on the incidence of tuberculosis in an Irish cattle population. *Preventive Veterinary Medicine* **34**, 47–56.

Otis, D.L., Burnham, K.P., White, G.C. and Anderson, D.R. (1978). Statistical inference from capture data on closed animal populations. *Wildlife Monographs* **62**, 1–135.

Owen-Smith, N. (2002). *Adaptive Herbivore Ecology. From Resources to Populations in Variable Environments.* Cambridge University Press: Cambridge.

Paine, R.T. (1966). Food web complexity and species diversity. *American Naturalist* **100**, 65–75.

Parkes, J.P. (1990a). Feral goat control in New Zealand. *Biological Conservation* **54**, 335–348.

Parkes, J.P. (1990b). Eradication of feral goats on islands and habitat islands. *Journal of the Royal Society of New Zealand* **20**, 297–304.

Parkes, J.P. (1993). The ecological dynamics of pest–resource–people systems. *New Zealand Journal of Zoology* **20**, 223–230.

Parkes, J.P., Norbury, G.L., Heyward, R.P. and Sullivan, G. (2002). Epidemiology of rabbit haemorrhagic disease (RHD) in the South Island, New Zealand, 1997–2001. *Wildlife Research* **29**, 543–555.

Paterson, R.A. (1990). Effects of long-term anti-shark measures on target and non-target species in Queensland, Australia. *Biological Conservation* **52**, 147–159.

Patterson, I.J., Abdul Jalil, S. and East, M.L. (1989). Damage to winter cereals by greylag and pink-footed geese in north-east Scotland. *Journal of Applied Ecology* **26**, 879–895.

Pavlov, P.M. and Hone, J. (1982). The behaviour of feral pigs, *Sus scrofa*, in flocks of lambing ewes. *Australian Wildlife Research* **9**, 101–109.

Pavlov, P.M., Hone, J., Kilgour, R.J. and Pedersen, H. (1981). Predation by feral pigs on Merino lambs at Nyngan, New South Wales. *Australian Journal of Experimental Agriculture and Animal Husbandry* **21**, 570–574.

Payton, I.J., Forester, L., Frampton, C.M. and Thomas, M.D. (1997). Response of selected tree species to culling of introduced Australian brushtail possums *Trichosurus vulpecula* at Waipaoua Forest, Northland, New Zealand. *Biological Conservation* **81**, 247–255.

Pech, R.P. and Hood, G.M. (1998). Foxes, rabbits, alternative prey and rabbit calicivirus disease: consequences of a new biological control agent for an outbreaking species in Australia. *Journal of Applied Ecology* **35**, 434–453.

Pech, R.P., Sinclair, A.R.E., Newsome, A.E. and Catling, P.C. (1992). Limits to predator regulation of rabbits in Australia: evidence from predator-removal experiments. *Oecologia* **89**, 102–112.

Pech, R.P., Hood, G.M., Singleton, G.R., Salmon, E., Forrester, R.I. and Brown, P.R. (1999). Models for predicting plagues of house mice (*Mus domesticus*) in Australia. In *Ecologically-based Management of Rodent Pests* (eds G.R. Singleton, L.A. Hinds, H. Leirs and Z. Zhang), pp. 81–112. Australian Centre for International Agricultural Research: Canberra.

Pekelharing, C.J., Parkes, J.P. and Barker, R.J. (1998). Possum (*Trichosurus vulpecula*) densities and impacts on fuchsia (*Fuchsia excorticata*) in South Westland, New Zealand. *New Zealand Journal of Ecology* **22**, 197–203.

Pimm, S.L. (1987). Determining the effects of introduced species. *Trends in Ecology and Evolution* **2**, 106–108.

Pimm, S.L. (1991). *The Balance of Nature?* Chicago University Press: Chicago.

Platt, J.R. (1964). Strong inference. *Science* **146**, 347–353.

Poche, R.M., Mian, Y., Haque, E. and Sultana, P. (1982). Rodent damage and burrowing characteristics in Bangladesh wheat fields. *Journal of Wildlife Management* **46**, 139–147.

Pople, A.R., Clancy, T.F., Thompson, J.A. and Boyd-Law, S. (1998). Aerial survey methodology and the cost of control for feral goats in western Queensland. *Wildlife Research* **25**, 393–407.

Potter, B. (1908). *The Tale of Jemima Puddle-Duck*. Frederick Warne & Co.: Harmondsworth.

Priddel, D. and Wheeler, R. (1990). Survival of malleefowl *Leipoa ocellata* chicks in the absence of ground-dwelling predators. *Emu* **90**, 81–87.

Priddel, D. and Wheeler, R. (1997). Efficacy of fox control in reducing the mortality of released captive-reared malleefowl, *Leipoa ocellata*. *Wildlife Research* **24**, 469–482.

Proulx, M. and Mazumder, A. (1998). Reversal of grazing impact on plant species richness in nutrient-poor vs. nutrient-rich ecosystems. *Ecology* **79**, 2581–2592.

Pulliam, H.R. (1988). Sources, sinks and population regulation. *American Naturalist* **132**, 652–661.

Putman, R.J. (1997). Deer and road traffic accidents: options for management. *Journal of Environmental Management* **51**, 43–57.

Raffaelli, D. and Moller, H. (2000). Manipulative field experiments in animal ecology: do they promise more than they deliver? *Advances in Ecological Research* **30**, 299–338.

Rainbolt, R.E. and Coblentz, B.E. (1997). A different perspective on eradication of vertebrate pests. *Wildlife Society Bulletin* **25**, 189–191.

Ralph, C.J. and Maxwell, B.D. (1984). Relative effects of human and feral hog disturbance on a wet forest in Hawaii. *Biological Conservation* **30**, 291–303.

Ramsey, D. (2005). Population dynamics of brushtail possums subject to fertility control. *Journal of Applied Ecology* **42**, 348–360.

Ramsey, D.S.L. and Engeman, R.M. (1994). Patterns of grazing on coastal dune systems by insular populations of two species of macropod. *Wildlife Research* **21**, 107–114.

Randolph, S.E., Chemini, C., Furlanello, C., Genchi, C., Hails, R.S., Hudson, P.J., Jones, L.D., Medley, G., Norman, R.A., Rizzoli, A.P., Smith, G. and Woolhouse, M.E.J. (2001). The ecology of tick-borne infections in wildlife reservoirs. In *The Ecology of Wildlife Diseases* (eds P.J. Hudson, A. Rizzoli, B.T. Grenfell, H. Heesterbeek and A.P. Dobson), pp. 119–138. Oxford University Press: Oxford.

Rasmussen, G.S.A. (1999). Livestock predation by the painted hunting dog *Lycaon pictus* in a cattle ranching region of Zimbabwe: a case study. *Biological Conservation* **88**, 133–139.

Redpath, S.M., Thirgood, S.J. and Leckie, F.M. (2001). Does supplementary feeding reduce predation of red grouse by hen harriers? *Journal of Applied Ecology* **38**, 1157–1168.

Ritchie, M.E. and Olff, H. (1999). Herbivore diversity and plant dynamics: compensatory and additive effects. In *Herbivores: Between Plants and Predators* (eds H. Olff, V.K. Brown and R.H. Drent), pp. 175–204. Blackwell Science: Oxford.

Robel, R.J., Dayton, A.D., Henderson, F.R., Meduna, R.L. and Spaeth, C.W. (1981). Relationships between husbandry methods and sheep losses to canine predators. *Journal of Wildlife Management* **45**, 894–911.

Rodda, G.H., Fritts, T.H. and Chiszar, D. (1997). The disappearance of Guam's wildlife. *Bioscience* **47**, 565–574.

Romesburg, H.C. (1981). Wildlife science: gaining reliable knowledge. *Journal of Wildlife Management* **45**, 293–313.

Romin, L.A. and Bissonette, J.A. (1996). Deer–vehicle collisions: status of state monitoring activities and mitigation efforts. *Wildlife Society Bulletin* **24**, 276–283.

Rosatte, R.C., Pybus, M.J. and Gunson, J.R. (1986). Population reduction as a factor in the control of skunk rabies in Alberta. *Journal of Wildlife Diseases* **22**, 459–467.

Rosatte, R.C., MacInnes, C.D., Williams, R.T. and Williams, O. (1997). A proactive prevention strategy for raccoon rabies in Ontario, Canada. *Wildlife Society Bulletin* **25**, 110–116.

Rosatte, R.C., Power, M.J., MacInnes, C.D. and Campbell, J.B. (1992). Trap-vaccinate-release and oral vaccination for rabies control in urban skunks, raccoons and foxes. *Journal of Wildlife Diseases* **28**, 562–571.

Ross, J.G., Hickling, G.J., Morgan, D.R. and Eason, C.T. (2000). The role of non-toxic prefeed and postfeed in the development and maintenance of 1080 bait shyness in captive brushtail possums. *Wildlife Research* **27**, 69–74.

Rowley, I. (1970). Lamb predation in Australia: incidence, predisposing conditions, and the identification of wounds. *CSIRO Wildlife Research* **15**, 79–123.

Rudolph, B.A., Porter, W.F. and Underwood, H.B. (2000). Evaluating immunocontraception for managing suburban white-tailed deer in Irondequoit, New York. *Journal of Wildlife Management* **64**, 463–473.

Ruscoe, W.A., Elkinton, J.S., Choquenot, D. and Allen, R.B. (2005). Predation of beech seed by mice: effects of numerical and functional responses. *Journal of Animal Ecology* **74**, 1005–1019.

Rutberg, A.T., Naugle, R.E., Thiele, L.A. and Liu, I.K.M. (2004). Effects of immunocontraception on a suburban population of white-tailed deer *Odocoileus virginianus*. *Biological Conservation* **116**, 243–250.

Sacks, B.N., Jaeger, M.M., Neale, J.C.C. and McCullough, D.R. (1999). Territoriality and breeding status of coyotes relative to sheep predation. *Journal of Wildlife Management* **63**, 593–605.

Saunders, G. (1988). The ecology and management of feral pigs in New South Wales. M.Sc. thesis. Macquarie University: Sydney.

Saunders, G. and Bryant, H. (1988). The evaluation of a feral pig eradication program during a simulated exotic disease outbreak. *Australian Wildlife Research* **15**, 73–81.

Savidge, J.A. (1987). Extinction of an island forest avifauna by an introduced snake. *Ecology* **68**, 660–668.

Schubert, C.A., Rosatte, R.C., MacInnes, C.D. and Nudds, T.D. (1998a). Rabies control: an adaptive management approach. *Journal of Wildlife Management* **62**, 622–629.

Schubert, C.A., Barker, I.K., Rosatte, R.C., MacInnes, C.D. and Nudds, T.D. (1998b). Effect of canine distemper on an urban raccoon population: an experiment. *Ecological Applications* **8**, 379–387.

Shea, K. and NCEAS Working Group on Population Management (1998). Management of populations in conservation, harvesting and control. *Trends in Ecology and Evolution* **13**, 371–375.

Sherman, P.W. and Morton, M.L. (1984). Demography of Belding's ground squirrels. *Ecology* **65**, 1617–1628.

Short, J. (1985). The functional response of kangaroos, sheep and rabbits in an arid grazing system. *Journal of Applied Ecology* **22**, 435–447.

Short, J., Bradshaw, S.D., Giles, J., Prince, R.I.T. and Wilson, G.R. (1992). Reintroduction of macropods (Marsupialia: Macropodoidea) in Australia – a review. *Biological Conservation* **62**, 189–204.

Sibly, R.M. and Hone, J. (2002). Determinants of population growth rate: an overview. *Philosophical Transactions of the Royal Society B* **357**, 1153–1170.

Sibly, R.M., Hone, J. and Clutton-Brock, T. (2003). *Wildlife Population Growth Rates*. Cambridge University Press: Cambridge.

Sinclair, A.R.E. (1996). Mammal populations: fluctuation, regulation, life history theory and their implications for conservation. In *Frontiers of Population Ecology* (eds R.B. Floyd, A.W. Shepherd and P.J. de Barro), pp. 127–154. CSIRO Publishing: Melbourne.

Sinclair, A.R.E., Pech, R.P., Dickman, C.R., Hik, D., Mahon, P. and Newsome, A.E. (1998). Predicting effects of predation on conservation of endangered prey. *Conservation Biology* **12**, 564–575.

Singleton, G.R., Hinds, L.A., Leirs, H. and Zhang, Z. (1999a). *Ecologically-based Management of Rodent Pests*. Australian Centre for International Agricultural Research: Canberra.

Singleton, G.R., Sudarmaji, J., Tan, T.Q. and Hung, N.Q. (1999b). Physical control of rats in developing countries. In *Ecologically-based Management of Rodent Pests* (eds G.R. Singleton, L.A. Hinds, H. Leirs and Z. Zhang), pp. 178–198. Australian Centre for International Agricultural Research: Canberra.

Singleton, G.R., Twigg, L.E., Weaver, K.E. and Kay, B.J. (1991). Evaluation of bromadiolone against house mouse (*Mus domesticus*) populations in irrigated soybean crops. II. Economics. *Wildlife Research* **18**, 275–283.

Skeat, A. (1990). Feral buffalo in Kakadu National Park: survey methods, population dynamics and control. M.App.Sci. thesis. University of Canberra: Canberra.

Smith, G.C., Cheeseman, C.L. and Clifton-Hadley, R.S. (1997). Modelling the control of bovine tuberculosis in badgers in England: culling and the release of lactating females. *Journal of Applied Ecology* **34**, 1375–1386.

Smith, G.C., Cheeseman, C.L., Wilkinson, D. and Clifton-Hadley, R.S. (2001). A model of bovine tuberculosis in the badger *Meles meles*: the inclusion of cattle and the use of a live test. *Journal of Applied Ecology* **38**, 520–535.

Smith, J.N.M., Taitt, M.J. and Zanette, L. (2002). Removing brown-headed cowbirds increases seasonal fecundity and population growth in song sparrows. *Ecology* **83**, 3037–3047.

Snedecor, G.W. and Cochran, W.G. (1967). *Statistical Methods*, 6th edn. Iowa State University Press: Ames.

Sol, D. and Senar, J.C. (1995). Urban pigeon populations: stability, home range, and the effect of removing individuals. *Canadian Journal of Zoology* **73**, 1154–1160.

Stahl, P., Vandel, J.M., Herrenschmidt, V. and Migot, P. (2001a). Predation on livestock by an expanding reintroduced lynx population: long-term trend and spatial variability. *Journal of Applied Ecology* **38**, 674–687.

Stahl, P., Vandel, J.M., Herrenschmidt, V. and Migot, P. (2001b). The effect of removing lynx in reducing attacks on sheep in the French Jura mountains. *Biological Conservation* **101**, 15–22.

Stahl, P., Vandel, J.M., Ruette, S., Coat, L., Coat, Y. and Balestra, L. (2002). Factors affecting lynx predation on sheep in the French Jura. *Journal of Applied Ecology* **39**, 204–216.

Stapp, P. (1998). A reevaluation of the role of prairie dogs in Great Plains grasslands. *Conservation Biology* **12**, 1253–1259.

Statham, M. (1994). Electric fencing for the control of wallaby movement. *Wildlife Research* **21**, 697–707.

Stenseth, N.C., Leirs, H., Skonhoft, A., Davis, S.A., Pech, R.P., Andreassen, H.P., Singleton, G.R., Lima, M., Machang'u, R.S., Makundi, R.H., Zhang, Z., Brown, P.R., Shi, D. and Wan, X. (2003). Mice, rats, and people: the bio-economics of agricultural rodent pests. *Frontiers in Ecology and Environment* **1**, 367–375.

Stephens, P.A. and Sutherland, W.J. (1999). Consequences of the Allee effect for behaviour, ecology and conservation. *Trends in Ecology and Evolution* **14**, 401–405.

Steury, T.D., Wirsing, A.J. and Murray, D.L. (2002). Using multiple treatment levels as a means of improving inference in wildlife research. *Journal of Wildlife Management* **66**, 292–299.

Stewart, W.B., Witmer, G.W. and Koehler, G.M. (1999). Black bear damage to forest stands in western Washington. *Western Journal of Applied Forestry* **14**, 128–131.

Stone, C.P. (1989). Native birds. In *Conservation Biology in Hawaii* (eds C.P. Stone and D.B. Stone), pp. 96–102. University of Hawaii Cooperative National Park Resources Studies Unit: Honolulu.

Stone, C.P. and Stone, D.B. (1989). *Conservation Biology in Hawaii.* University of Hawaii Cooperative National Park Resources Studies Unit: Honolulu.

Sukumar, R. (1991). The management of large mammals in relation to male strategies and conflict with people. *Biological Conservation* **55**, 93–102.

Summers, R.W. (1990). The effect on winter wheat of grazing by brent geese *Branta bernicula*. *Journal of Applied Ecology* **27**, 821–833.

Swinton, J., Woolhouse, M.E.J., Begon, M.E., Dobson, A.P., Ferroglio, E., Grenfell, B.T., Guberti, V., Hails, R.S., Heesterbeek, J.A.P., Lavazza, A., Roberts, M.G., White, P.J. and Wilson, K. (2001). Microparasite transmission and persistence. In *The Ecology of Wildlife Diseases* (eds P.J. Hudson, A. Rizzoli, B.T. Grenfell, H. Heesterbeek and A.P. Dobson), pp. 83–101. Oxford University Press: Oxford.

Sykes, J.B. (1976). *The Concise Oxford Dictionary of Current English*, 6th edn. Clarendon Press: Oxford.

Tapper, S.C., Potts, G.R. and Brockless, M.H. (1996). The effect of an experimental reduction in predation pressure on the breeding success and population density of grey partridges *Perdix perdix*. *Journal of Applied Ecology* **33**, 965–978.

Tchamba, M.N. (1996). History and present status of the human/elephant conflict in the Waza-Logone Region, Cameroon, West Africa. *Biological Conservation* **75**, 35–41.

Thomas, C.D. and Kunin, W.E. (1999). The spatial structure of populations. *Journal of Animal Ecology* **68**, 647–657.

Thorne, E.T. and Williams, E.S. (1988). Disease and endangered species: the black-footed ferret as a recent example. *Conservation Biology* **2**, 66–74.

Thouless, C.R. and Sakwa, J. (1995). Shocking elephants: fences and crop raiders in Laikipia District, Kenya. *Biological Conservation* **72**, 99–107.

Tilman, D. and Lehman, C.L. (1997). Habitat destruction and species extinctions. In *Spatial Ecology. The Role of Space in Population Dynamics and Interspecific Interactions* (eds D. Tilman and P. Karieva), pp. 233–249. Princeton University Press: Princeton.

Tisdell, C.A. (1982). *Wild Pigs: Environmental Pest or Economic Resource?* Pergamon Press: Sydney.

Tobin, M.E., Koehler, A.E., Sugihara, R.T., Ueunten, G.R. and Yamaguchi, A.M. (1993). Effects of trapping on rat populations and subsequent damage and yields of macadamia nuts. *Crop Protection* **12**, 243–248.

Tobin, M.E., Koehler, A.E. and Sugihara, R.T. (1997). Effects of simulated rat damage on yields of macadamia trees. *Crop Protection* **16**, 203–208.

Trout, R.C., Ross, J., Tittensor, A.M. and Fox, A.P. (1992). The effect on a British wild rabbit population (*Oryctolagus cuniculus*) of manipulating myxomatosis. *Journal of Applied Ecology* **29**, 679–686.

Turner, M.G. (1988). Simulation and management implications of feral horse grazing on Cumberland Island, Georgia. *Journal of Range Management* **41**, 441–447.

Tuyttens, F.A.M., Barron, L., Rogers, L.M., Mallinson, P.J. and Macdonald, D.W. (2000). Wildlife management and scientific research: a retrospective evaluation of two badger removal operations for the control of bovine tuberculosis. In *Mustelids in a Modern World. Management and Conservation Aspects of Small Carnivores: Human Interactions* (ed. H.I. Griffiths), pp. 247–265. Backhuys Publishers: Leiden.

Twigg, L.E. and Williams, C.K. (1999). Fertility control of overabundant species; can it work for feral rabbits? *Ecology Letters* **2**, 281–285.

Twigg, L.E., Singleton, G.R. and Kay, B.J. (1991). Evaluation of bromadiolone against house mouse (*Mus domesticus*) populations in irrigated soybean crops. I. Efficacy of control. *Wildlife Research* **18**, 265–274.

Twigg, L.E., Lowe, T.J., Martin, G.R., Wheeler, A.G., Gray, G.S., Griffin, S.L., O'Reilly, C.M., Robinson, D.J. and Hubach, P.H. (2000). Effects of surgically imposed sterility on free-ranging rabbit populations. *Journal of Applied Ecology* **37**, 16–39.

Twigg, L.E., Martin, G.R. and Lowe, T.J. (2002). Evidence of pesticide resistance in medium-sized mammalian pests: a case study with 1080 poison and Australian rabbits. *Journal of Applied Ecology* **39**, 549–560.

Ujvari, M., Baagoe, H.J. and Madsen, A.B. (1998). Effectiveness of wildlife warning reflectors in reducing deer–vehicle collisions: a behavioral study. *Journal of Wildlife Management* **62**, 1094–1099.

Underwood, A. (1997). *The Design of Experiments*. Cambridge University Press: Cambridge.

van den Bosch, F., Metz, J.A.J. and Diekmann, O. (1990). The velocity of spatial population expansion. *Journal of Mathematical Biology* **28**, 529–565.

VanDruff, L.W., Bolen, E.G. and San Julian, G.J. (1994). Management of urban wildlife. In *Research and Management Techniques for Wildlife and Habitats*, 5th edn (ed. T.A. Bookhout), pp. 507–530. Wildlife Society: Bethesda.

van Tets, G.F. (1969). Quantitative and qualitative changes in habitat and avifauna at Sydney airport. *CSIRO Wildlife Research* **14**, 117–128.

VerCauteren, K.C., Dolbeer, R.A. and Gese, E.M. (2005). Identification and management of wildlife damage. In *Techniques for Wildlife Investigations and Management*, 6th edn (ed. C.E. Braun), pp. 740–778. Wildlife Society: Bethesda.

Vtorov, I.P. (1993). Feral pig removal: effects on soil microarthropods in a Hawaiian rain forest. *Journal of Wildlife Management* **57**, 875–880.

Vucetich, J.A. and Waite, T.A. (2000). Is one migrant per generation sufficient for the genetic management of fluctuating populations? *Animal Conservation* **3**, 261–266.

Wagner, K.K. and Conover, M.R. (1999). Effect of preventive coyote hunting on sheep losses to coyote predation. *Journal of Wildlife Management* **63**, 606–612.

Walker, B. (1998). The art and science of wildlife management. *Wildlife Research* **25**, 1–9.

Walker, B.H. and Norton, G.A. (1982). Applied ecology: towards a positive approach. II. Applied ecological analysis. *Journal of Environmental Management* **14**, 325–342.

Walsh, B. and Whitehead, P.J. (1993). Problem crocodiles, *Crocodylus porosus*, at Nhulunbuy, Northern Territory: an assessment of relocation as a management strategy. *Wildlife Research* **20**, 127–135.

Walter, M. (2002). The population ecology of wild horses in the Australian alps. PhD thesis. University of Canberra: Canberra.

Walters, C. (1986). *Adaptive Management of Renewable Resources*. Macmillan: New York.

Warburton, B. and Drew, K.W. (1994). Extent and nature of cyanide-shyness in some populations of Australian brushtail possums in New Zealand. *Wildlife Research* **21**, 599–605.

Warne, R.M. and Jones, D.N. (2003). Evidence of target specificity in attacks by Australian magpies on humans. *Wildlife Research* **30**, 265–267.

Whisson, D.A., Orloff, S.B. and Lancaster, D.L. (1999). Alfalfa yield loss from Belding's ground squirrels in northeastern California. *Wildlife Society Bulletin* **27**, 178–183.

White, G. (1775). *Gilbert White's Year. Passages from The Garden Kalendar and the Naturalist's Journal*. Oxford University Press: Oxford.

White, J., Horskins, K. and Wilson, J. (1998). The control of rodent damage in Australian macadamia orchards by manipulation of adjacent non-crop habitats. *Crop Protection* **17**, 353–357.

White, P.C.L., Groves, H.L., Savery, J.R., Conington, J. and Hutchings, M.R. (2000). Fox predation as a cause of lamb mortality on hill farms. *Veterinary Record* **147**, 33–37.

Williams, C.K. and Moore, R.J. (1995). Effectiveness and cost-efficiency of control of the wild rabbit, *Oryctolagus cuniculus* (L.), by combinations of poisoning, ripping, fumigation and maintenance fumigation. *Wildlife Research* **22**, 253–269.

Williams, J.R., Nokes, D.J., Medley, G.F. and Anderson, R.M. (1996). The transmission dynamics of hepatitis B in the U.K.: a mathematical model for evaluating costs and effectiveness of immunization programmes. *Epidemiology and Infection* **116**, 71–89.

Wilson, E.O. (1992). *The Diversity of Life*. Penguin Books: London.

Wilson, K., Bjornstad, O.N., Dobson, A.P., Merler, S., Poglayen, G., Randolph, S.E., Read, A.F. and Skorping, A. (2001). Heterogeneities in macroparasite infections: patterns and processes. In *The Ecology of Wildlife Diseases* (eds P.J. Hudson, A. Rizzoli, B.T. Grenfell, H. Heesterbeek and A.P. Dobson), pp. 6–44. Oxford University Press: Oxford.

Wintner, S.P. and Dudley, S.F.J. (2000). Age and growth estimates for the tiger shark, *Galeocerdo cuvier*, from the east coast of South Africa. *Marine and Freshwater Research* **51**, 43–53.

Wobeser, G.A. (1994). *Investigation and Management of Disease in Wild Animals*. Plenum Press: New York.

Wood, B.J. and Fee, C.G. (2003). A critical review of the development of rat control in Malaysian agriculture since the 1960s. *Crop Protection* **22**, 445–461.

Wywialowski, A.P. (1996). Wildlife damage to field corn in 1993. *Wildlife Society Bulletin* **24**, 264–271.

Yom-Tov, Y., Ashkenazi, S. and Viner, O. (1995). Cattle predation by the golden jackal *Canis aureus* in the Golan Heights, Israel. *Biological Conservation* **73**, 19–22.

Zhang, Z., Chen, Z., Ning, Z. and Huang, X. (1999). Rodent pest management in agricultural ecosystems in China. In *Ecologically-based Management of Rodent Pests* (eds G.R. Singleton, L.A. Hinds, H. Leirs and Z. Zhang), pp. 261–284. Australian Centre for International Agricultural Research: Canberra.

Index